T0212037

ASTRONOMY AND
ASTROPHYSICS LIBRARY

For further volumes:
http://www.springer.com/series/848

A. Satya Narayanan

An Introduction to Waves and Oscillations in the Sun

 Springer

A. Satya Narayanan
Indian Institute of Astrophysics
Bangalore, India

ISSN 0941-7834
ISBN 978-1-4614-4399-5 (hardcover)   ISBN 978-1-4614-4400-8 (eBook)
ISBN 978-1-4899-9596-4 (softcover)
DOI 10.1007/978-1-4614-4400-8
Springer New York Heidelberg Dordrecht London

Library of Congress Control Number: 2012946177

Dedicated to
Professor Bernard Roberts (Bernie)
With Respectful Regards

Preface

Waves and oscillations are present everywhere. In particular, the Sun, which is the nearest star (whose disk is clearly visible), compared to the other stars in the Milky Way and those in other galaxies, has a variety of waves and oscillations. The classification of these waves depends on the external forces acting on the Sun. It is a natural plasma laboratory, in which both experimental and theoretical studies of plasma can be applied and verified with observations. The most important waves that are present in the Sun are the Alfvén wave and the fast and slow magnetoacoustic wave, which arise due to compressibility effects.

The first chapter provides a brief introduction to the Sun and its structure and composition, such as density, pressure, temperature, and other plasma parameters. Also dealt with are the different features and their morphology. The second chapter introduces basic ideas of electromagnetics, to pave the way for the subsequent discussions on magnetohydrodynamics (MHD). In addition to the waves and oscillations present in the Sun, there are other dynamic phenomena taking place in the Sun. A brief discussion of the basic concepts of MHD, its equations, and assumptions is introduced. Some simple analytic solutions of the complicated MHD equations, under simplified, yet physical situations, are discussed. Notably, the concepts of flux tubes, current-free and force-free magnetic fields, a simple model of the prominences, and the relationship between the vorticity and the induction equation are mentioned in passing. Finally, the Parker solution, which describes the phenomenon of solar wind, is introduced.

Chapters 4 and 5 deal with the theoretical aspects of waves and oscillations in homogeneous and nonhomogeneous structured media. Chapter 4 has more relevance to Alfvén and sound waves. The effect of gravity, shear flows, is mentioned briefly. Observational signatures and nonlinear studies on waves in homogeneous media are mentioned briefly. Chapter 5, which deals with waves in a nonuniform media, discusses waves in interfaces (magnetic, density discontinuities), uniform slab, and cylindrical geometries. A brief introduction to waves in an annulus and twisted magnetic flux tube is included in this chapter.

Waves and oscillations, in general, exhibit instabilities. Theoretical studies of instabilities are rather difficult. However, a brief introduction to well-known instabilities, such as the Rayleigh–Taylor instability, Kelvin–Helmholtz instability, and parametric instability, is presented in Chap. 6. Also, the magnetic buoyancy (Parker) instability and its importance in astrophysical flows are included.

The importance of the waves and oscillations present in the Sun from an observational point of view is reviewed in Chap. 7. There are discussions of waves in sunspots, the 5-min oscillations, chromospheric oscillations, and oscillations in the corona. An introduction to Moreton and EIT (Extreme ultraviolet Imaging Telescope) waves is included at the end of the chapter. Chapter 8 deals with helioseismology, a branch of solar physics. This method helps in getting a clear picture of the internal structure of the Sun, based on the analysis of the several modes (global) of oscillations present in the Sun.

Bangalore, India   A. Satya Narayanan

Acknowledgments

This book includes some of the works by the author. However, the immense contribution of several experts working in the field has helped the author greatly. He has had personal interaction with some and knows the others through their work. The author wishes to acknowledge the following people: E. R. Priest, B. Roberts, M. Goossens, J. L. Ballester, M. S. Ruderman, R. Erdelyi, V. M. Nakariakov, M. J. Aschwanden, J. C. Dalsgard, T. Sakurai, N. Gopalswamy, K. Somasundaram, S. S. Hasan, P. Venkatakrishnan, B. N. Dwivedi, H. M. Antia, C. Uberoi. He has also utilized figures generated by some of these authors for this book. To those who are not mentioned explicitly, he renders his apology.

The author expresses his sincere thanks to his colleagues, R. K. Chaudhuri, M. V. Mekkaden, A. V. Raveendran, and S. K. Saha, who encouraged him in this endeavor and gave him confidence.

Special thanks are due to R. Ramesh, who went through the whole manuscript, for his positive criticisms and suggestions. Baba Varghese, C. Kathiravan, and Indrajit Barve are thanked for their help in generating diagrams and images.

The author wishes to convey his heartfelt regard to Dr. Sreepat Jain, whose encouragement and efforts made the book possible. He is grateful to the reviewers for the positive recommendations and suggestions. Many thanks to Maury Solomon and Megan Ernst of Springer for their initiative and support. The publisher, Springer, has done a wonderful job, and the author is thankful to them.

Finally, no author can hope to write a book without the support of the family. The author had the able support of his wife (Sukanya) and son (Prahladh S. Iyer). To both of them, he owes a lot.

Contents

Chapter 1
Introduction

1.1 Historical Perspectives

AUM BHOOR BHUWAH SWAHA,
TAT SAVITUR VARENYAM
BHARGO DEVASAYA DHEEMAHI
DHIYO YO NAHA PRACHODAYAT (Sanskrit)

Translation:

Oh God! Thou art the Giver of Life,
Remover of pain and sorrow,
The Bestower of happiness,
Oh! Creator of the Universe,
May we receive thy supreme sin-destroying light,
May Thou guide our intellect in the right direction

A Prayer to the "Giver of Light and Life"—the Sun (Savitur)

The above verse has been included to emphasize the fact that civilized man had realized centuries ago the importance of the Sun, our closest star, for his everyday livelihood. Initially, he started to worship the Sun as God. Throughout human history, the view of the Sun has been one of special reverence, both practical and mystical. Cultures around the globe have revered the Sun as a divine being and utilized the predictability of its annual wanderings through the stars to mark special times of the year. Many early records, such as Egyptian sun temples, aboriginal star lore, Native American medicine wheels, and many other astronomical structures around the world, mark special points in the Sun's apparent path in the sky (for example, the solstices, equinoxes, and eclipses). There were some individuals (like us) who were not content with worshiping the Sun as God. They observed that the Sun did not rise in the same position every day, observed the different phases of the Moon, from the full moon to the new moon phase, and that the phase from the

A. Satya Narayanan, *An Introduction to Waves and Oscillations in the Sun*, Astronomy and Astrophysics Library, DOI 10.1007/978-1-4614-4400-8_1,
© Springer Science+Business Media New York 2013

full moon to the new moon was periodic. They watched eclipses, both of the Sun and the Moon. Initially, they interpreted it as some demon, swallowing them.

We can now boldly make the statement that we will have no home without the Sun. The amount of energy provided by the Sun, being the closest star, has set the conditions for the formation of life on Earth, which circles around the star some 150 million kilometers from the center of the solar system. We still do not know if there are signs of life on other planets (similar to ours), although we do know for sure that there are more than 300 planets (typically like Jupiter and Saturn) discovered in other planetary systems. The life forms, if they exist at all in other exo-planets, may or may not be similar to the ones on the Earth. Thus, for the time being, let's assume that the Sun is unique, that it has provided us enough opportunity to evolve.

Man's impression about the stars, planets circling around him, was that he was at the center of the universe (the geocentric universe). When he realized that the Sun did not rise in the same place every day, he started contemplating the reasons behind it. He was not equipped with the ideas of relative motion, curvature of the Earth, and the rotation of the Moon, the Sun, and other celestial objects. The idea of an Earth-centered universe was supported by observation; Ptolemy, who lived in the early part of the second century, developed a model in which the planets wheeled around on numerous interlocking circular paths, known as epicycles, which provided the backbone of the geocentric worldview. Such a conclusion was due to the result of the dictates of Aristotle and his teacher Plato, who lived in the fourth century BC. Their argument was that the Earth was at the center of the universe and that everything in the heavens was perfect and therefore unchanging, moving at constant speed in perfect circles. In earlier days, a circle was believed to be the perfect geometrical figure. To match the detailed observations, Ptolemy had to resort to a complicated overlap of circular pathways. However, with the publication of the book *On the Revolutions of the Heavenly Spheres*, by the Polish astronomer Nicolaus Copernicus in 1543, the geocentric model of the solar system had to be replaced by the heliocentric model ("helio" means "sun" in Greek).

The ideas developed by Copernicus led to a revolution of thought by the German mathematician and astronomer Johannes Kepler (1571–1630), who believed that the heliocentric idea proposed by Copernicus was more apt and probable. After decades of struggling to understand the motion of the planets, Kepler eventually made a revolutionary breakthrough. This was made possible by the careful and painstaking observations made by his teacher, Tycho Brahe. He threw out the ideas of Aristotelian constraints of circular orbits and constant velocities. He solved the problem by assuming that the planets moved in elliptical orbits, contrary to the conventional belief that they are circles. Kepler's laws of planetary motion are presented here briefly: (1) The orbit of each planet around the Sun is an ellipse, with the Sun at one of the focii; (2) the planet moves around its orbit in such a way that it sweeps equal areas at equal intervals of time; (3) the cube of the average distance of the planet from the Sun is proportional to the square of its orbital period, that is, $p^2 = a^3$. Here, p is the orbital period and a is the average distance. The above laws, based entirely on observations, paved the way for Isaac Newton's theory of gravitation.

The concepts of time, calendar, and so forth evolved from the careful observation of the path the Sun traced every day for one full year. From our perspective on the Earth, the Sun travels across the sky from the east to west, with the shape and location of the path changing every day over the year. Actually, this motion is caused by the Earth's eastward rotation, causing the Sun to rise in the east and set in the west. The Earth revolves around the Sun with its axis tilted 23.5° relative to the plane that contains the Sun and the planets (the ecliptic plane) (Sonnet et al. 1991). In particular, the tilt of the Earth's axis is important, as this is the main reason for the changing seasons experienced on the Earth. The location of the rising of the Sun throughout the year, if one observes carefully, tends to change, from farther north as the year progresses from winter to summer (in the Northern Hemisphere), while the Sun rises south of due east in the winter and north of due east in the summer. The Sun rises due east on the equinoxes, summer (March 20/21) and spring (September 22/23). The tilt in the Earth's axis creates four important latitudes, namely, the Arctic and Antarctic circles, and the tropics of Cancer and Capricorn.

The modern calendar originates from the dates of the ancient Egyptians, who used the 365-day year as early as 4200 BC. Their calendar was based on the time between the spring equinoxes (the tropical year), wherein the Sun crosses the equator on its way north. Since the tropical year in principle is actually 365.2422 days long, their calendar would drift by one day every four years with respect to the seasonal changes. After a period of 100 years, the equinox would occur almost a month earlier. This led to the concept of the leap year, where a day was added once in four years to match the calendar with the seasonal variations. The concept of a 24 h/day is linked to the passage of the Sun. The length of a day is defined as the time the planet takes to rotate once about its axis. Thus, the concept of a day is different for different planets. The Earth rotates in 23 h 56 min and 4.1 s (to be precise), which is slightly less than 24 h. The 24 h is based on the time it takes the Sun to cross the sky, namely, the time between when the Sun is at its highest point in the sky on successive days. This is defined as the solar day and is 24 h on average, with a variation of $+/-25$ s. The additional motion of the Earth around the Sun leads to the difference between the solar day and the rotational period of the Earth. It is interesting to note that since the tropical year is 365.242375 days, in order for the spring equinox to occur on the same time every year, leap days are not added to a new century, unless it is divisible by 400 and not 4. This makes up for the difference of 1/100 between 365 and 365.242375 days. For example, 2400 will be a leap year, while 2500 will not be one.

From now on, we will discuss the Sun, its evolution, other physical characteristics, such as its density, its pressure, the magnetic field, the distance from the Earth, and so on, from a scientific point of view. The different patterns exhibited by the Sun are very important, although the focus is changing considerably. The assumption that the Sun is a perfect unchanging heavenly body has changed permanently. This was possible due to the efforts of Galileo (1564–1642), who turned his newly invented telescope toward the Sun and identified the sunspots as being solar in nature. The change in the position of the sunspots across the disk of the Sun, their changing numbers, and sizes only demonstrated the fact that the Sun is a variable star on many different scales, which led to the modern solar physics (Foukal 2004). In the present century, advances in technology for global communications, navigation, and weather monitoring make us susceptible to the vagaries of the Sun's behavior (Strong et al. 1998). The modern approach to observing the Sun has a lot to do with the pattern and growth of the sunspots over the solar cycle (an 11-year period), along with the production of the energetic activity.

The study of astrophysical objects in general relies on the detailed knowledge of how physical systems interact to produce light and other electromagnetic radiation, and the Sun is no exception to it. While the Sun, which is approximately 150 million kilometers away, is very close in astronomical time scales, compared to any other star (the next closest being the proxima centauri), it is not close enough for us to directly measure the physical characteristics of its constituent parts. (Clark 2007; Finze 2008). The luxury and facility to stick a thermometer into the solar atmosphere to measure its temperature or to check by shining light through to see how much of it is absorbed or transmitted are far from being a reality. Sitting on Earth, we can only measure whatever radiation comes to us from the Sun. In some special cases, we may detect it from space through instruments on board the different spacecrafts (scientific) that have been launched. Most of the information we have about the Sun comes from the radiation we receive and observe, be it the X-ray photons, the million-degree corona, energetic particles accelerated by solar storms, or the optical light from the Sun's surface. The sharp edge of the Sun that we see in telescopes is referred to as the surface of the Sun by the scientists working on the Sun.

Ever since Galileo discovered the sunspots (the dark regions), we have gathered several interesting features on the Sun over the years. However, a number of mysteries still remain. We have yet to understand the Sun's outermost atmosphere, that is, the hot corona and the million-degree K, temperature it possesses. This is much higher than the visible photosphere, which has a temperature of about 6,000 K. The violent release of energy in the form of flares from the chromosphere (a thin layer of the solar atmosphere located in between the photosphere and the corona) has yet to be resolved completely. Acceleration of particles to very high energies, sending large volumes of coronal material hurtling out into space at several thousand kilometers per second, the so-called coronal mass ejection (CME), the generation of the Sun's magnetic field in the interior, making its way into the surface and beyond, where it dominates the solar activity, are some of the challenges for which we have some answers, while the complete story has yet to be finished. The regular solar cycle gets interrupted sometimes, resulting in a reduction in the numbers of sunspots, with noticeable climatic effects on the Earth.

The Sun is the nearest star, at a distance of 1.5×10^{11} m from us, and the source of life on the Earth. It is a relatively mediocre, middle-aged star with many great features, stationed in the neighborhood of the outer spiral arm of the spiral galaxy known as the Milky Way (Lang 2006; Giovanelli 1984). It lies approximately 30,000 light-years from the galactic center and takes about 250 million years to go around the galaxy. It is a burning ball of incandescent gas, 860,000 miles in diameter. The Sun produces heat and light by thermonuclear reactions taking place inside the core.

The study of the Sun as a star, and indeed the study of the stars in general, started with the invention of the spectrograph as an observational tool. The word "spectroscopy" is derived from the word "spectrum," which is the terminology used to explain the continuous spread of colors caused by the dispersion of light by a prism. The real importance of spectroscopy came about after the work of the German physicist Joseph Fraunhofer (1787–1826) in the early 1800s. He saw that sunlight, when passed through the spectrograph, was broken up into a large number of dark lines (now known as Fraunhofer lines). These were the first spectral lines ever observed and allowed great advances to be made in the science of spectroscopy, and our knowledge about stars and the Sun, in particular, increased considerably. However, Fraunhofer did not understand what caused them. It was the key observation made by the French physicist Jean Foucault, who observed that a flame containing the element sodium would absorb the yellow light emitted by a bright arc formed between two carbon electrodes. However, the laboratory experiments revealed that different sources emitted bright lines of varying colors (emission lines), while Foucault's observation demonstrated for the first time that dark lines at varying wavelengths (absorption lines) could be generated. The correct interpretation of these observations was made by the famous physicists Gustav Kirchhoff (1824–1887) and Robert Bunsen (1811–1899). They showed that each gas had its own unique spectrum. Kirchhoff was able to show that at a given wavelength, the power emitted and the power absorbed are the same for all objects at the same temperature (the famous Kirchhoff law). The Fraunhofer lines in the solar spectrum could now be explained as the absorption of specific wavelengths by the different elements making up the solar surface. A careful analysis of these absorption lines made the study of the atmosphere of the Sun as well as other stars possible. It is a well-known fact that today stars are distinguished by their spectral type. The Sun is known as a yellow main sequence star of spectral type G2.

Using Newton's law of gravitation, one can calculate the observable quantities, the mass of the Sun, once the relationship between the orbital period and the distance is known. The square of the period of a planet's orbit is proportional to the cube of the distance to the Sun divided by the mass of the Sun (the famous law due to Kepler). Knowing the period in seconds and the distance in kilometers, one can calculate the mass of the Sun, which turns out to be approximately 2×10^{30} kg. The age of some of the oldest objects in the solar system, meteorites, is used as a proxy for determining the age of the Sun. Some theories assert that the solar system collapsed from a single gaseous nebula, with the Sun forming slightly ahead of the planets, asteroids, and comets. The method of decay of the radioactive isotope of Rubidium-87, which decays into the stable isotope Strontium-87, is used to determine the ages of the meteorites. The age of the Sun is close to 4.57 billion years (Table 1.1).

The Sun as a star plays a major role in understanding the stars and astronomical universe. Proximity reveals details of its surface at widely ranging spatial and temporal scales that are not possible for any other star. Many physical processes that occur elsewhere in the universe can be examined in detail on the Sun. It should be obvious to us that the Sun is a star because it shines. The Sun produces its

Table 1.1 Physical
characteristics of the Sun

Properties	Values
Radius (km)	696,000
Mass (kg)	1.988×10^{30}
Volume (m^3)	1.41×10^{27}
Average density (kg/m^2)	1,408
Surface gravity (m/s^2)	273.95
Rotation period (days)	26 (at equator)
Temperature at surface (K)	5,785
Escape velocity at surface (km/h)	2,223 million

own energy, in the form of heat and light, which is radiated to the solar system. A well-known law of nature is that the energy of a system can neither be created nor destroyed. This is the law of conservation of energy. However, energy can be converted from one state to another state. For example, when we rub our hands, they get warm—we convert biochemical energy in the arm muscles received from food and drink to the kinetic energy in the motion of the hands, which in turn transfers the thermal energy via friction between the hands. This clearly shows that in a similar way, the Sun produces heat and light, radiant energy by some energy conversion processes. Discovering how the Sun and other stars produce the vast amount of energy required to keep them burning for at least 5 billion years was one of the fundamental landmarks of the previous century. For almost a century, the source of the Sun's energy was a great topic of debate among scientists. By the end of the nineteenth century, the distance to the Sun had been calculated reasonably accurately, which enabled us to calculate the physical size of the Sun. Using measurements of the energy output of the Sun at the Earth, astronomers were able to calculate the amount of radiant energy emitted by the Sun. For example, the Sun produces enough energy in one second to power the entire United States for several centuries. This vast amount of energy has to be accounted for by some plausible energy conversion process. Many suggestions and calculations have been made by astronomers through the ages. The true nature was discovered after scientists realized the power of nuclear reactions with Einstein's famous equation, which gave the relationship between mass and energy by the simple formula $E = mc^2$. The only source of energy sufficient enough to produce this large power is the energy released in the fusion of atoms at the core of the Sun.

The structure and properties of the outer, visible atmosphere of the Sun are investigated through spectroscopic studies. The Sun is a magnetic star. Magnetic fields are observed through tracers and measured by Zeeman effect (Priest 1982a). The atmosphere of the Sun is turbulent, and a variety of features are observed on it, including granulation, oscillation, fibrils, spicules, transient brightenings, coronal holes, and solar winds (corona). Mechanisms responsible for active and explosive events (observed mainly in the chromosphere and transition region), dynamic loops, such as a violent solar energy release (flares, CMEs), which are magnetically driven, are still not fully understood.

The first scientific theories involved chemical reactions or gravitational collapse. Chemical burning has been ruled out, as it cannot account for the Sun's luminosity. The conversion of gravitational potential energy into heat as the Sun contracts would only keep the Sun shining for 25 million years. Late nineteenth-century geological research indicated the Earth was older than the Sun. However, the development of nuclear physics led to the correct answer. Given the mass of the Sun, this will provide enough energy for the Sun to shine for 10 billion years. The Sun began as a cloud of gas undergoing gravitational collapse. The same heating process, once proposed to power the Sun, did cause the core of the Sun to get hot and dense enough to start nuclear fusion reactions. Once begun, the fusion reactions generated energy, which provided an outward pressure. This pressure perfectly balances the inward force of gravity. Deep inside the Sun, the pressure is strongest when gravity is strong, while near the surface, the pressure is weakest when gravity is weaker. This balance is called gravitational equilibrium, which causes the Sun's size to remain stable.

Modern theories state that the Sun and the solar system formed from the gravitational collapse of a large cloud of gas known as the solar nebula. Most of the stars are born from nebulae (a thick cloud of gaseous matter) (Aller 1998). As the cloud collapsed, the central region that was later to form the Sun got denser and denser, which resulted in the heating, ultimately forming a large ball of hot gas with a very dense and hot core. It is elementary physics that a compression of a hot gas increases the internal heat of the gas. The intense heat in the core was so powerful that the atoms there, mostly hydrogen, could not maintain their structure and were broken up into free protons and electrons that moved around quickly, as they had a lot of thermal energy. The core of the Sun is also a region of high density, so that fast-moving particles cannot go very far before they hit other particles moving just as fast. An interesting observation at this juncture is the fact that light (photons) reaches the earth in 8 light minutes, traveling a distance of about 150 million kilometers, while it takes several hundreds of years to start from the core and reach the surface of the Sun. Thus, one can imagine the density of matter that is prevalent inside the Sun.

The Sun consists mainly of hydrogen and helium. One may wonder if the radiation from the core arises mainly from the fusion of hydrogen nuclei into helium nuclei. The answer is far from being simple. Two hydrogen nuclei (H) collide. A positron (e^+) and a neutrino (N) are produced. The properties of the positron are similar to those of an electron (which is negative), but it has a positive charge. A deuterium (D) nucleus remains. When this collides with a proton, a nucleus of a helium isotope ^{3}He is produced, which consists of two protons and one neutron. Getting two protons to stick together is not that easy, as they both have a positive charge and like charges repel. However, in the core , the protons are moving so fast and collide with such violence that they can overcome the repulsive force of the similar charges to get close enough to stick together via a force that particle physicists call the "strong force." If we try to push two magnets of like poles gently, they deflect. Push them hard and exert enough force to overcome the natural repulsion, and they will touch each other. Protons coming together will result in a loss in the form of radiation. Overall, this chain of reactions has fused four hydrogen nuclei into one helium nucleus. The energy that is thus released gives rise to the Sun's radiation. Normally, mass does not spontaneously convert to energy. However, in the extreme conditions at the core of a star, the violent interactions between the atomic nuclei cause mass and energy to be converted in the form of radiation. It is well known that helium-4 ion is 0.7% less massive than four protons. This 0.7% of mass that goes astray in the production of helium from hydrogen is not lost, but is converted into energy. In all, roughly 600 billion kg of hydrogen are fused into helium every second in the center of the Sun. The mass converted to energy is then simply 0.7% of this mass, which is approximately 4.2 billion kg. By using Einstein's energy equation, one can calculate and realize that the Sun's core produces approximately 3.78×10^{26} J of energy. The Sun is made up of mostly hydrogen and helium, with 70% and 28%, respectively. The remaining 2% accounts for oxygen, carbon, and iron. We know this by identifying the absorption lines in the Sun's spectrum. These lines are formed in the photosphere. If we analyze starlight, we can find a star's temperature, chemical composition, and radial velocity (Doppler). A 100- km-thick layer of hot gases, with T 6,000 K, emits 99.99% of energy generated in the solar interior, mostly in the visible spectral range centered at 5,000 Å. High-atmosphere structures are rooted in the photosphere/subphotosphere. The composition, temperature, and pressure of the photosphere are revealed by the dark Fraunhofer absorption lines, over 20,000 identified in the spectrum of sunlight (Chaplin 2006).

A brief introduction to the electromagnetic spectrum will be very useful at this stage. Electromagnetic radiation is that which carries energy and moves through vacuous space in periodic waves at the speed of light, propagated by the interplay of oscillating electric and magnetic fields. The velocity of light is usually designated by the letter "c" and has a value of 299792.458 km/s. Electromagnetic radiation includes radio waves, infrared radiation, visible or optical radiation, ultraviolet radiation, X-rays, and gamma rays. Electromagnetic radiation, in common with any wave, has a wavelength, denoted by λ, and a frequency, denoted by ν; their product is equal to the velocity of light, or $\lambda \nu = c$, so the wavelength decreases when the frequency increases, and vice versa. The energy associated with the radiation increases in direct proportion to frequency, and this energy, known as the photon energy and denoted by E, is given by $E = h\nu = hc/\lambda$, where Planck's constant $h = 6.6261 \times 10^{-34}$ J s. There is a continuum of electromagnetic radiation—from long-wavelength radio waves of low frequency and energy, through visible-light waves, to gamma rays of higher frequency and energy. All types of wavelengths of electromagnetic radiation mentioned above will represent the electromagnetic energy spectrum. An example of how the sizes vary from the subatomic level to the size of buildings is presented in Fig. 1.1. This figure also shows the penetration frequency of the electromagnetic spectrum at different frequencies. Also shown are the corresponding temperatures as a function of the wavelength. It is very clear from the figure that the atmosphere is transparent to radio and visible wavelengths, while observing the Sun at other wavelengths is possible by instruments flown on board spacecrafts.

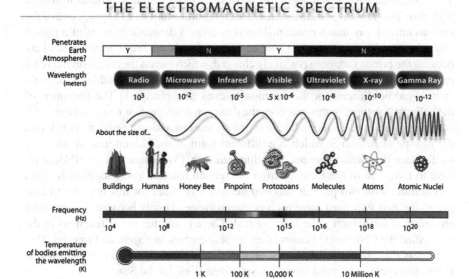

Fig. 1.1 The electromagnetic spectrum; from NASA, USA

1.2   The Core of the Sun

The core is the innermost 10% of the Sun's mass. It is where the energy from nuclear fusion is generated. The core of the Sun is a gigantic nuclear cauldron, in which atoms of hydrogen are cooked into atoms of helium. Because of the enormous amount of gravity compression from all of the layers above it, the core is very hot and dense. The difference in the mass is turned into energy, which appears at the surface of the Sun as light and heat (Hufbauer 1993; Kippenhahn 1994). The temperature at the very center of the Sun is about 14,500,000°C; it has a density around 160 times the density of water. This is over 20 times denser than the dense metal iron, which has a density "only" 7 times that of water. Nuclear fusion requires extremely high temperatures and densities. Both the temperature and the density decrease as one moves outward from the center of the Sun. The nuclear burning is almost completely shut off beyond 0.3 solar radii (about 27,175,000 km from the center). At that point, the temperature is only half the central value, and the density drops to about 20 g/cc. However, the Sun's interior is still gaseous all the way to the very center because of the extreme temperatures. There is no molten rock like that found in the interior of the Earth. As mentioned earlier, the main process taking place in the core of the Sun is the nuclear burning. The Sun exists because of a process called fusion or the proton cycle. In the core of the Sun, the temperature is so hot that the atoms are constantly colliding and tearing apart the hydrogen atoms to form separate protons, neutrons, and electrons. These are the parts that make up an atom. The "freeing" of these particles is what lets the proton cycle take place.

There are three basic reactions in this process: (1) The fusion of protons to form a deuteron: Two protons collide in this reaction. In order for this to happen, one of the protons must decay into a neutron. This is because a deuteron consists of a proton and a neutron. The protons must also be very close to each other, which is unlikely because the protons are both positively charged, which means they are likely to repel each other. It's a good thing that there are SO many protons available in the core, or it would be difficult for this reaction to even take place! (2) The formation of the helium-3 isotope: In this reaction, the deuteron from the first reaction fuses with another proton to make a combination of two protons and one neutron. This forms the isotope of helium-3, which is a different form of the helium that we are used to. To make that helium, we need the third reaction. (3) Helium is formed! Here we have to have two of the isotopes from the second reaction come together to form real helium, with two protons and two neutrons. In order for this reaction to take place, the first two must have each occurred twice. This is because two isotopes are required for helium to be fully formed. There is another set of reactions in the Sun called the CNO cycle because they involve carbon, nitrogen, and oxygen. This cycle was thought to be the main source of the Sun's energy. Today we know that the proton cycle is the most important set of reactions for the Sun.

1.3   Radiative Zone

The radiative zone is a region of highly ionized gas. There the energy transfer is primarily by photon diffusion. The radiative zone is where the energy is transported from the super-hot interior to the colder outer layers by photons. Technically, this also includes the core. The radiative zone includes the inner approximately 85% of the Sun's radius. Radiation is a very important aspect of the Sun. It is the main process through which the Sun transfers its energy out into space. The radiative zone of the solar interior is characterized by the process of radiation (see Fig. 1.2). The energy made in the core is in the form of photons, more specifically in gamma rays, when it first begins its journey outward. This energy is changed into less energetic photons as it moves through the radiative zone. This is good for us because gamma rays are very dangerous to humans! In radiation, energy diffuses out from the core through these photons. They move very quickly (at the speed of light!), but they also bounce off so many other particles that it takes hundreds of thousands of years for them to get through the radiative zone. All of the bouncing off of other particles sends the photons flying off in all directions instead of taking a straight path outward. This is called a "random walk." The energy generated in the core is carried by light photons that bounce from particle to particle through the radiative zone. Although the photons travel at the speed of light, they bounce so many times through this dense material that an individual photon takes about a million years to finally reach the interface layer. The density drops from 20 g/cc to 0.2 g/cc from the bottom to the top of the radiation zone. The temperature falls from 7,000,000°C to

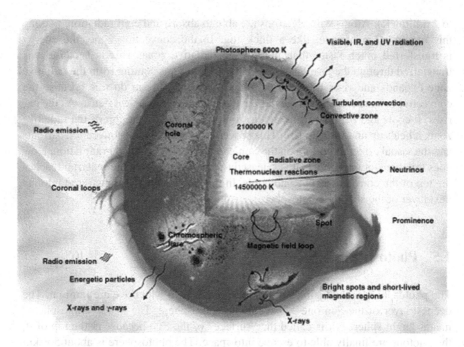

Photosphere 6000 K

Visible, IR, and UV radiation

Turbulent convection

Convective zone

Radio emission

Coronal hole

2100000 K

Core   Radiative zone

Thermonuclear reactions

Neutrinos

14500000 K

Coronal loops

Spot

Prominence

Chromospheric flare

Magnetic field loop

Radio emission

Energetic particles

Bright spots and short-lived magnetic regions

X-rays and γ-rays

X-rays

Fig. 1.2 The structure of the Sun; from NASA, USA

2,000,000°C over the large distance. Radiation moving out from this part of the Sun is absorbed more readily, reducing the amount that actually makes its way out of the Sun. This makes the gas unstable and leads to convection.

1.4 Convection Zone

The convection zone is the outermost layer of the solar interior. It extends from a depth of about 200,000 km right up to the visible surface. At the base of the convection zone, the temperature is about 2,000,000°C. The convective motions carry heat quite rapidly to the surface. The fluid expands and cools as it leaves. At the visible surface, the temperature has dropped to 5,700°C, and the density is only 0.0000002 g/cc. Energy in the outer 15% of the Sun's radius is transported by the bulk motions of gas in a process called convection. At cooler temperatures, more ions are able to block the outward flow of photon radiation more effectively, so nature kicks in convection to help the transport of energy from the very hot interior to the cold space. This is the area that we consider to form the outer shell of the Sun. The atoms in this layer of the Sun have electrons because the temperature is not hot enough to strip them away like it is in the core (15.6×10^6 K as opposed to 2 million K). Atoms with electrons are able to absorb and emit radiation, making this region more opaque, like a thick fog. In the convection zone, the energy is transferred much faster than it is in the radiative zone. This is because it is transferred through the process of convection. Hotter gas coming from the radiative zone expands and rises through the convective zone. It can do this because the convective zone is cooler than the radiative zone and therefore less dense. As the gas rises, it cools and begins to sink again. As it falls down to the top of the radiative zone, it heats up and starts to rise. This process repeats, creating convection currents and the visual effect of boiling on the Sun's surface. This is called granulation. How does this transfer energy? Heat is released to the outside when the material reaches the top of the convective zone and cools. In this way, energy is transferred into the next layer of the sun, the photosphere.

1.5 Photosphere

The photosphere is the visible surface of the Sun with which we are familiar. The deepest layer of the Sun one can see is the photosphere. The word "photosphere" means "light sphere." It is called the "surface" of the Sun because at the top of it, the photons are finally able to escape into space. The photosphere is about 500 km thick. Remember that the Sun is totally gaseous, so the surface is not something you could land or float on. It is a dense enough gas that you cannot see through it. It emits a continuous spectrum. Several methods of measuring the temperature have all determined that the Sun's photosphere has a temperature of about 5,840 K. Although we refer to the photosphere as a layer of the Sun, in actuality it is a part of the Sun's atmosphere (Durrant 1988). It is a very thin layer in comparison with the rest of the Sun and is the only part of the Sun that we can actually see when looking at it from Earth, because the photosphere is where the light is emitted. (But, of course, you should never look straight at the Sun!) The light that we see coming from the Sun is actually far from its real intensity because the photosphere's opaqueness absorbs much of it. So the place that creates the light absorbs it as well! Looking at different places on the Sun changes where the visible light comes from. When looking at the center of the disk, the light that we see comes from the base of the photosphere. But as we look closer to the limb, the light comes from higher up, so at the very edge of the Sun, it is emerging from a spot far above the base of the photosphere. This makes the Sun look less bright and slightly redder on the limb, and brighter on the disk. In the photosphere, granulation, super granulation, faculae, and sunspots are seen.

Sunspots: Galileo discovered that the Sun's surface is sprinkled with small dark regions called sunspots. These are seen as dark spots in the photosphere that have extremely high magnetic fields. They usually show up in groups of two sets, where one set has a north magnetic field and the other set has a south magnetic field (Thomas and Weiss 2008). Sunspots are cooler regions on the photosphere. Since they are 1,000–1,500 K cooler than the rest of the photosphere, they do not emit as much light and appear darker. They can last a few days to a few months. Galileo used the longer-lasting sunspots to map the rotation patterns of the Sun. Because the Sun is gaseous, not all parts of it rotate at the same rate. The solar equator rotates once every 25 days, while regions at 30° above and below the equator take 26.5 days to rotate, and regions at 60° from the equator take up to 30 days to rotate. They have a lower temperature than their surroundings, which gives them this darkened appearance in white light. Hundreds of years of observing the sunspots on the Sun have shown that the number of sunspots varies in a cycle with an average period of 11 years. At the start of a sunspot cycle, the number of sunspots is at a minimum and most of them are within ±30° from the solar equator. At solar maximum, about 5.5 years later when the sunspot number peaks, most of the sunspots are within just 5° of the solar equator. Sunspots are regions of strong magnetic fields. This affects the spectral lines in the sunspot spectra. Each absorption line will split up into multiple components. The amount of separation between the components measures the strength of the magnetic field. The magnetic field is somehow responsible for the sunspot cycle. In one 11-year cycle, the leading sunspot in a sunspot group will have a north magnetic pole, while the trailing sunspot in the group will have a south magnetic pole. In the next 11-year cycle, the poles will switch, and so the total cycle is 22 years long. Sunspots form where twisted magnetic field lines rise out of the photosphere and then loop back down into the photosphere and deeper layers. The magnetic field lines suppress the convection at those points on the photosphere, so energy has a harder time leaking out at those points on the photosphere—they are cooler than the rest of the photosphere. In the chromosphere above sunspots, there are structures called prominences. These prominences are bright clouds of gas that follow the magnetic field lines. So-called quiet prominences form in the corona (the Sun's atmosphere) about 40,000 km above the surface. They form loops of hydrogen gas (as the gas follows the loops) in the magnetic field. Quiet prominences last several days to several weeks. "Surge" prominences lasting up to a few hours shoot gas up to 300,000 km above the photosphere.

Faculae: Faculae are seen on the Sun near its limb (the edge of the photosphere). Instead of appearing dark like sunspots, they show up as bright spots on the photosphere. This is because they are hotter than their surroundings. They are magnetic also, but their magnetism is more concentrated than that of sunspots.

Granulation: Granules are related to the convective zone. The granulation that shows up in the photosphere is a result of the rising and falling of hot gas that takes place in the convective zone. The bubbles seen are the material that reaches the top of the convective zone—the photosphere (Bray et al. 1984).

Super granulation: Super granules are just larger version of granules. They have magnetic field "bunches" that flow within them. Super granules look similar to granules, except that they are 35,000 km across as opposed to 1,000 km across.

1.6 Chromosphere

The chromosphere is a narrow layer above the photosphere that rises in temperature with height. The chromosphere is an irregular layer above the photosphere where the temperature rises from 6,000 to 20,000°C. At these temperatures, hydrogen emits light that gives off a reddish color. The chromosphere is also the site of activity as well. Changes in polar solar flares, prominences, and filament eruption, and the flow of material in postflare loops can be observed over the chromosphere. Normally, it can't be seen by the naked eye because the light from the photosphere of the Sun overpowers it. However, during a solar eclipse when this light is blocked out, the chromosphere appears as a narrow, red ring around the Sun, with an irregular outer edge. The light from the chromosphere is also visible in prominences when they project from the Sun. The edge of the chromosphere is made up of spicules. These are narrow columns of material that ascend into the corona and last about 15 min. They are smaller eruptions but eject material into the corona at high speeds. A lot of other solar events also take place within the chromosphere, such as solar flares and prominences. Solar flares (Tandberg-Hansen and Emslie 1988) are a catastrophic enhancement of energy over the entire electromagnetic spectrum, from radio waves at the long wavelength end, through optical emission, to X-rays and gamma rays at the short wavelength end, with particle acceleration occurring over a localized area on the atmosphere of the Sun. A solar flare occurs when the magnetic energy that builds up in the solar atmosphere is suddenly released. The energy release is in the range of 10^{21}–10^{26} J in a few minutes to several hours in minor to major events. One of the features of the chromosphere is the chromospheric network. It outlines the super granules (see the photosphere section for an explanation) and is present there because of the magnetic field bunches in the super granules. The network makes a web pattern of magnetic field lines on the Sun. One of the interesting things about the chromosphere is the way in which its temperature rises with height. One would expect that the temperature would decrease as the radiation coming from the photosphere moves up and more energy leaks out into space. This means there must be some other form of energy present that has nothing to do with the radiation coming from below the chromosphere. Scientists believe that this source of heating deals with wave motions, specifically magnetohydrodynamic waves. They are created when a magnetic field line is displaced. When the line tries to go back to its original shape, it begins to oscillate. These oscillations create waves that give up energy as they move through plasma and cause the strange rise of temperature in the chromosphere.

1.7 Corona

Above the chromosphere, a pearly white halo called the corona extends tens of millions of kilometers into space. The corona is continually expanding into the interplanetary space and in this form is called the "solar wind". When the new moon covers up the photosphere during a total solar eclipse, one can see the pearly white corona around the dark moon (Golub and Pasachoff 2010). This is the rarefied upper atmosphere of the Sun. It has a very high temperature of 1 to 2 million K. Despite its high temperature, it has a low amount of heat because it is so tenuous. The temperature of the corona is hotter than 1,000,000°C, while the visible surface has a temperature of about 6,000°C. The nature of the processes that heat the corona, maintain at these high temperatures, and accelerate the solar wind is a great mystery in solar physics. The corona is known to be very hot because it has ions with many electrons removed from the atoms (the plasma state). At high enough temperatures, the atoms collide with each other with enough energy to eject electrons. This process is called "ionization." At very high temperatures, atoms like iron can have 9–13 electrons ejected. Nine-times ionized iron is only produced at temperatures of 1.3 million K, and 13-times ionized iron means the temperature gets up to 2.3 million K! During strong solar activity, the temperature can reach 3.6 million K and lines from 14-times ionized calcium are seen. The corona is the collection of immediate gases around the Sun. It is extremely hot, much hotter than the surface of the Sun. Like the chromosphere, it can be seen with a naked eye during a solar eclipse, as this is the only time that the light from the photosphere is blocked out enough so that anything else can be seen. It can also be observed with a coronagraph, which is an instrument that can produce an artificial eclipse that blocks out light from the photosphere. The coronal light is just the scattered light from the photosphere, which is why its color is the same of that of the photosphere. Most of the corona is trapped close to Sun by loops of magnetic field lines (called coronal loops). In X-rays, the corona appear bright. Some magnetic field lines do not loop back to the Sun and will appear dark in X-rays. These places are called "coronal holes." Helmet coronal streamers emit from the Sun in long, pointed, funnel-shaped structures. They usually arise from sunspots and active regions, so at the base of a helmet streamer, one will often find a prominence. They form magnetic loops that connect the sunspots and suspend material above the surface of the Sun. The magnetic field lines trap the material to form the streamers.

Coronal holes: Coronal holes are regions where the corona is dark. They are often found at the Sun's poles and are associated with open magnetic field lines. Most of the solar wind originates from these holes in the corona. They can only be seen by looking at the Sun through an X-ray telescope. Coronagraphs allow us to see the corona on the limb, but in order to see it on the disk, it has to be looked at through an X-ray telescope. X-rays allow us to see things with high temperatures, and since the photosphere is cool, the X-ray telescope blocks out the light from it so that the corona can be seen around the disk. Being able to look at the corona in this way reveals its structure. It shows that it consists of loops and arches of material that originate from the chromosphere and photosphere. Some of these loops are associated with transient solar flares, while others last longer. These loops are denser than their surroundings. Several mechanisms have been suggested as the source of heating, but there is no consensus on which one, or combination thereof, is actually responsible.

Coronal mass ejections (CME): The most dramatic temporal evolution in the corona occurs in coronal mass ejection events, CMEs, which in turn produce the largest transient disturbances in the solar wind. The speed of a typical CME will be of the order of 500 km/s, although for extreme cases, the speed can go up to 2,000 km/s or more. The fastest CME can have a kinetic energy of about 10^{26} J. The shock ahead of a fast CME is broader than the CME that drives it. The ambient magnetic field drapes about the CME. In the normal solar wind, field lines are open to the outer boundary of the heliosphere, and a single field-aligned, anti-sunward-directed strahl is observed (Crooker et al. 1997). CMEs originate in closed field regions in the corona, and field lines within CMEs are at least initially connected to the Sun at both ends. Counterstreaming strahls are commonly observed on closed field lines and help identify CMEs in the solar wind (ICMEs). Every CME carries a new magnetic flux into the heliosphere. A magnetic reconnection in the foot points serves to open up the closed field loops associated with the CME, produces helical field lines within the CME, and helps to maintain a roughly constant magnetic flux in the heliosphere.

1.8   Solar Wind

The solar wind is a plasma, that is, an ionized gas, that fills the solar system. It results from the supersonic expansion of the solar corona (see Fig. 1.3). The solar wind consists primarily of electrons and protons with a smattering of alpha particles and other ionic species at low abundance levels. At 1 AU (Earth), the average proton densities, flow speeds, and temperatures are $8.7 \, \text{cm}^{-3}$, 468 km/s, and 1.2×10^5 K, respectively. Embedded within the solar wind is a magnetic field having an average strength of 6.2 nano-tesla at 1 AU. The solar wind plays an essential role in shaping and stimulating planetary magnetospheres and ionic comet tails. It is a prime source of space weather. Carrington's 1859 observation of a white light solar flare, followed 17 h later by a large geomagnetic storm, suggested a possible cause and effect. Lindemann (early 1900s) suggested large geomagnetic storms resulted from interactions between plasma clouds ejected from the Sun during flares and the Earth's magnetic field. Observations of recurrent (at 27-day rotation period of Sun) geomagnetic storms led to a hypothesis of M (for magnetic) regions on the Sun that produced long-lived streams of charged particles in interplanetary space. There almost always is at least a low level of geomagnetic activity. This suggested that plasma from the Sun is always present near the Earth. Observations by S. Forbush in the 1930s and 1940s of modulations of cosmic rays in association with geomagnetic storms and in association with an 11-year solar activity cycle suggested modulations were caused by magnetic fields embedded in plasma clouds from the Sun. Biermann concluded in the early 1950s that a continuous outflow of particles from the Sun filling interplanetary space was required to explain the anti-sunward orientation of ionic comet tails.

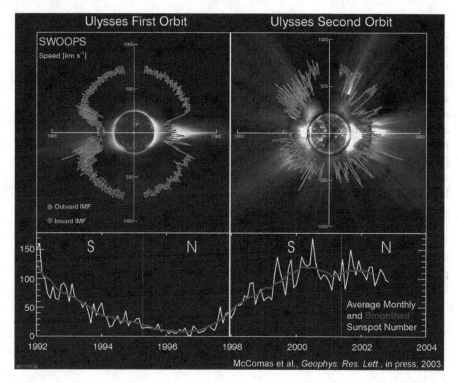

Fig. 1.3 Ulysses picture of the solar wind; from NASA, USA

In 1958, motivated by diverse indirect observations, E. N. Parker developed the first fluid model of a continuously expanding solar corona driven by the large pressure difference between the solar corona and the interstellar plasma. His model produced low flow speeds close to the Sun, supersonic flow speeds far from the Sun, and vanishingly low pressures at large heliocentric distances. In view of the fluid character of the model, he called this continuous supersonic expansion the solar wind. The electrical conductivity of the solar wind plasma is so high that the solar magnetic field is frozen into the solar wind flow as it expands outward from the Sun. Because the Sun rotates with a period of 27 days as observed from the Earth, magnetic field lines in the Sun's equatorial plane are bent into spirals whose inclination to the radial direction depend on the heliocentric distance and the speed of the wind. At 1 AU, the average field is inclined ≈45° to the radial direction in the equatorial plane. Measurements made by an electrostatic analyzer and a magnetometer on board Mariner II during its epic three-month journey to Venus in 1962 provided firm confirmation of a continuous solar wind flow and spiral heliospheric magnetic field that agree with Parker's model, on average. Mariner II also showed that the solar wind is highly variable, being structured into alternating streams of high- and low-speed flows that last for several days each. The observed magnetic field was also highly variable in both strength and orientation, while the solar rotation produces radial variations in speed. Faster wind overtakes slow wind ahead while outrunning slow wind behind. As a result, the leading edges of high-speed streams steepen with increasing heliocentric distance. Plasma is compressed on the leading edge of a stream and rarefied on the trailing edge. The buildup of pressure on the leading edge of a stream produces forces that accelerate the low-speed wind ahead and decelerate the high-speed wind within the stream. When the difference in speed between the crest of a stream and the trough ahead is greater than about twice the sound speed, ordinary pressure signals do not propagate fast enough to move the slow wind out of the path of the fast wind, and a forward-reverse shock pair forms on the opposite sides of the high-pressure region. Although the shocks propagate in opposite directions relative to the solar wind, both are carried away from the Sun by the high-bulk flow of the wind. The major accelerations and decelerations of the wind then occur at the shocks and the stream profile becomes a damped double sawtooth. Because the sound speed decreases with increasing heliocentric distance, virtually all high-speed streams eventually have shock pairs on their leading edges. The dominant structure in the solar equatorial plane in the outer heliosphere is the expanding compression regions, where most of the plasma and magnetic field are concentrated.

Chapter 2
Electromagneto Statics

2.1 Charge and Current Distributions

In the previous chapter, we discussed the physical properties of the Sun, such as mass, density, temperature, and distance from the Earth. We also briefly touched upon the different layers of the Sun, from the core, which is the innermost part of the Sun, up to the outermost atmosphere, namely, the solar wind. To understand the physical processes underlying the evolution of the Sun, one needs to look at the different dynamic processes taking place inside and outside the surface of the Sun. Whenever the dynamics of any physical system is to be understood, one has to carefully study the different forces that influence the system. One of the fundamental forces that influences the Sun and other heavenly bodies is the gravitational force.

We all know that electric and magnetic forces play an important role in various physical situations. However, depending on the nature of the problem of interest, one tends to ignore the effect of the electric field over the magnetic field, and vice versa. For example, the magnetic fields play a very crucial role in the evolution of certain features, such as flares, prominences, coronal mass ejection, and so forth, in the solar atmosphere described in the previous chapter. Given the situation, it would be important at this juncture to introduce some basic concepts of both electro- and magnetostatics and the Maxwell equations of electromagnetism. The theory of magnetohydrodynamics, a branch of plasma physics, with the continuum hypothesis and Maxwell's and fluid equations, is a useful tool to understand the dynamical features observed in the Sun.

The detection of charge: The presence of an electric charge on a body can be detected only by the forces that the electric charge causes the body to produce or experience. In other words, it is difficult to tell by simply looking at a particular object whether it is charged. The only option is to check whether it can produce or experience an electrostatic force (Manners 2000).

Types of charges: There are mainly two types of charges. Bodies that carry the same kind of charge repel one another, whereas bodies carrying different types of charge

A. Satya Narayanan, *An Introduction to Waves and Oscillations in the Sun*, Astronomy and Astrophysics Library, DOI 10.1007/978-1-4614-4400-8_2,
© Springer Science+Business Media New York 2013

attract one another. A body carrying one type of charge can become electrically neutral (does not exert any electric forces) by absorbing an equal quantity of the other type of charge. This property of charge cancellation has led to the two types of charge being labeled positive $(+)$ and negative $(-)$ because the sum of an amount of positive charge and an equal amount of negative charge is zero.

Source of electric charge: Matter generally contains an electric charge. Atoms consist of a nucleus, which is made up of protons, neutrons, and electrons, which orbit the nucleus. Electrons carry a negative charge and protons a positive charge, while the neutron is electrically neutral (as the name suggests). When two different materials are brought into contact, electrons may be transferred from one material to the other. The direction of transfer of the electrons depends on the properties of the materials concerned and is always the same for any two materials.

Conservation of charge: When two materials come into contact with each other, due to friction, the existing charges are simply redistributed between the two materials. Thus, the total amount of charge in any isolated system is also a constant. The conservation of charge is similar to the laws of conservation of linear momentum, angular momentum, and energy as one of the fundamental laws of physics. However, the conservation of charge does not imply that charges can never be created nor destroyed. Instead, it implies that for any positive charge created, an equal amount of negative charge must also appear.

Conduction of charge: Bodies carrying unlike charges attract one another, while any body that has a deficit of electrons (i.e., any body that carries a net positive charge) will not only attract negatively charged macroscopic bodies in the vicinity, but will also attract any electrons that are close by. If these electrons are free to move, they will flow toward the positively charged body and neutralize it. Materials that allow an electric charge to flow through them are called conductors and those that do not are called insulators. An introduction to the above concepts may be found in the following books: Greiner (1998), Schwartz (1972), Ulaby (1997).

Charge densities: In electromagnetic theory, one encounters various forms of electric charge distributions, and if the charges are in motion, they constitute current distributions. A charge may be distributed over a volume of space, across a surface, or along a line. At the atomic scale, the charge distribution in a material is discrete, that is, the charge exists only where electrons and nuclei are, and nowhere else. In electromagnetics, we are interested in studying phenomena at larger scales, typically three or more orders of magnitude greater than the spacing between adjacent atoms. For such a macroscopic scale, we ignore the discontinuous nature of the charge distribution and treat the net charge contained in an elemental volume $\triangle V$ as if it were uniformly distributed within it. We can define the volume charge density ρ_v as

$$\rho_V = \text{limit}_{\triangle V \to 0} \frac{\triangle q}{\triangle V} = \frac{dq}{dV}, \tag{2.1}$$

where $\triangle q$ is the charge contained in $\triangle V$. In general, ρ_V is defined at a given point in space, specified by (x, y, z) in a Cartesian coordinate system, and at a given time t, that is, $\rho_V = \rho_V(x,y,z)$. Physically, ρ_V represents the average charge per unit volume for a volume $\triangle V$ centered at (x,y,z), with $\triangle V$ being large enough to contain a large number of atoms and yet small enough to be regarded as a point at the macroscopic scale under consideration. The variation of ρ_V with spatial location is called its spatial distribution, or simply its distribution. The total charge contained in a given volume V is given by

$$Q = \int \rho_V dV. \tag{2.2}$$

In some cases, particularly when dealing with conductors, the electric charge may be distributed across the surface of a material, in which case the relevant quantity of interest is the surface charge density ρ_s, defined as

$$\rho_s = \mathrm{limit}_{\triangle s \to 0} \frac{\triangle q}{\triangle s} = \frac{dq}{ds}, \tag{2.3}$$

where $\triangle q$ is the charge present across an elemental area $\triangle s$. Similarly, if the charge is distributed along a line, which need not be straight, we can characterize the distribution in terms of the line charge density ρ_l, defined as

$$\rho_l = \mathrm{limit}_{\triangle l \to 0} \frac{\triangle q}{\triangle l} = \frac{dq}{dl}. \tag{2.4}$$

Consider a tube of charge with volume charge density ρ_V in which the charges are moving with a mean velocity $\mathbf{u}$ along the axis of the tube. Over a period $\triangle t$, the charges move a distance $\triangle l = u \triangle t$. The amount of charge that crosses a cross-sectional surface $\triangle s'$ of the tube in time $\triangle t$ is therefore

$$\triangle q' = \rho_V \triangle V = \rho_V \triangle l \triangle s' = \rho_V u \triangle s' \triangle t. \tag{2.5}$$

Consider the more general case where the charges are flowing through a surface $\triangle s$ whose surface normal $\hat{\mathbf{n}}$ is not necessarily parallel to $\mathbf{u}$. In this case, the amount of charge $\triangle q$ flowing through $\triangle s$ is given by

$$\triangle q = \rho_V \mathbf{u} \cdot \triangle \mathbf{s} \triangle t, \tag{2.6}$$

and the corresponding current is

$$\triangle I = \frac{\triangle q}{\triangle t} = \rho_V \mathbf{u} \cdot \triangle \mathbf{s} = \mathbf{J} \cdot \triangle \mathbf{s}, \tag{2.7}$$

where

$$\mathbf{J} = \rho_V \mathbf{u} \tag{2.8}$$

is defined as the current density. For an arbitrary surface S, the total current flowing through it is then given by

$$I = \int_S \mathbf{J} \cdot d\mathbf{s}. \tag{2.9}$$

Consider the force between two very small charged bodies (i.e., bodies whose diameters are very small compared to the distance between them). This allows one to assume that all the charge on each of the bodies was concentrated at a point, which is often referred to as a point charge. In the late eighteenth century, Coulomb was able to show that the magnitude of the electrostatic force acting on two charges q_1 and q_2 was (1) inversely proportional to r^2 when the magnitudes of q_1 and q_2 were fixed, (2) proportional to the magnitude of q_1 when q_2 and r were fixed, and (3) proportional to the magnitude of q_2 when q_1 and r were fixed. When combined, these give the single relationship

$$F \propto \frac{|q_1||q_2|}{4\pi\varepsilon_0 r^2}. \tag{2.10}$$

The quantity ε_0 is known as the permittivity of free space.

2.2 Coulomb's Law

Coulomb's law was initially introduced for electrical charges in air and later generalized to material media. It states that an isolated charge q induces an electric field $\mathbf{E}$ at every point in space, and at any specific point P, it is given by

$$\mathbf{E} = \hat{\mathbf{R}} \frac{q}{4\pi\varepsilon R^2}, \tag{2.11}$$

where $\hat{\mathbf{R}}$ is a unit vector pointing from q to P, R is the distance between them, and ε is the electrical permittivity of the medium containing the observation point P. The expression given by Eq. (2.11) for the field $\mathbf{E}$ due to a single charge can be extended to find the field due to multiple point charges. Consider two point charges q_1 and q_2, located at position vectors $\mathbf{R}_1$ and $\mathbf{R}_2$ from the origin of a given coordinate system. The electric field $\mathbf{E}$ is to be evaluated at a point P with position vector $\mathbf{R}$. At P, the electric field $\mathbf{E}_1$ due to q_1 is given by Eq. (2.11) with R, the distance between q_1 and P, replaced with $|\mathbf{R} - \mathbf{R}_1|$ and the unit vector $\hat{\mathbf{R}}$ replaced with $(\mathbf{R} - \mathbf{R}_1)/|\mathbf{R} - \mathbf{R}_1|$. Thus,

$$\mathbf{E}_1 = \frac{q_1(\mathbf{R} - \mathbf{R}_1)}{4\pi\varepsilon|\mathbf{R} - \mathbf{R}_1|^3}. \tag{2.12}$$

Similarly, the electric field due to q_2 is

$$\mathbf{E}_2 = \frac{q_2(\mathbf{R} - \mathbf{R}_2)}{4\pi\varepsilon|\mathbf{R} - \mathbf{R}_2|^3}. \tag{2.13}$$

The electric field obeys the principle of linear superposition. Consequently, the total electric field $\mathbf{E}$ at any point in space is equal to the vector sum of the electric fields induced by all the individual charges. In the present case,

$$\mathbf{E} = \mathbf{E}_1 + \mathbf{E}_2 = \frac{1}{4\pi\varepsilon} \left[\frac{q_1(\mathbf{R} - \mathbf{R}_1)}{|\mathbf{R} - \mathbf{R}_1|^3} + \frac{q_2(\mathbf{R} - \mathbf{R}_2)}{|\mathbf{R} - \mathbf{R}_2|^3} \right]. \tag{2.14}$$

Generalizing the preceding result to the case of N point charges, the electric field $\mathbf{E}$ at position vector $\mathbf{R}$ caused by charges $q_1, q_2, \ldots, q_N$, located at points with position vectors $\mathbf{R}_1, \mathbf{R}_2, \ldots, \mathbf{R}_N$, is given by

$$\mathbf{E} = \frac{1}{4\pi\varepsilon} \sum_{i=1}^{N} \frac{q_i(\mathbf{R} - \mathbf{R}_i)}{|\mathbf{R} - \mathbf{R}_i|^3}. \tag{2.15}$$

2.3 Gauss's Law

The electric field for a single point charge q, situated at the origin and radius r, can also be written as

$$\mathbf{E}(\mathbf{r}) = \frac{1}{4\pi\varepsilon} \frac{q}{r^2} \hat{\mathbf{r}}. \tag{2.16}$$

For the point charge q at the origin, the flux of $\mathbf{E}$ through a sphere of radius r is

$$\int \mathbf{E} \cdot d\mathbf{a} = \int \frac{1}{4\pi\varepsilon} \left(\frac{q}{r^2} \hat{\mathbf{r}} \right) \cdot (r^2 \sin\theta \, d\theta \, d\phi \, \hat{\mathbf{r}}) = \frac{q}{\varepsilon}. \tag{2.17}$$

According to the principle of superposition, the total field is the (vector) sum of all the individual fields:

$$\mathbf{E} = \sum_{j=1}^{N} \mathbf{E}_i. \tag{2.18}$$

The flux through a surface that encloses them all is then given by

$$\int \mathbf{E} \cdot d\mathbf{a} = \sum_{i=1}^{N} \left(\int \mathbf{E}_i \cdot d\mathbf{a} \right) = \sum_{i=1}^{N} \left(\frac{1}{\varepsilon} q_i \right). \tag{2.19}$$

For any closed surface, then

$$\int_S \mathbf{E} \cdot d\mathbf{a} = \frac{1}{\varepsilon} Q_{\text{enc}}, \tag{2.20}$$

where Q_{enc} is the total charge enclosed within the surface. This is the quantitative statement of Gauss's law. As it stands, Gauss's law is an integral equation, but we can readily turn it into a differential one by applying the divergence theorem:

$$\int_S \mathbf{E} \cdot d\mathbf{a} = \int_V (\nabla \cdot \mathbf{E}) d\tau. \qquad (2.21)$$

Rewriting Q_{enc} in terms of the charge density ρ, we have

$$Q_{enc} = \int_V \rho \, d\tau. \qquad (2.22)$$

Thus, Gauss's law becomes

$$\int_V (\nabla \cdot \mathbf{E}) d\tau = \int_V \left(\frac{\rho}{\varepsilon}\right) d\tau. \qquad (2.23)$$

This holds for any volume. Thus, the integrands must be equal:

$$\nabla \cdot \mathbf{E} = \frac{1}{\varepsilon} \rho. \qquad (2.24)$$

The above equation is Gauss's law in differential form. While the differential form is simple, the integral form has the advantage that it accommodates point, line, and surface charges more naturally. An important property of Gauss's law is that for symmetrical cases, Gauss's law provides one of the quickest and easiest means of calculating electric fields. Gauss's law has the disadvantage that it is not useful for nonsymmetric cases. Three cases of symmetry that one encounters in physics are (1) plane symmetry, (2) cylindrical symmetry, and (3) spherical symmetry.

Electric potential: The interesting property of the electric field $\mathbf{E}$ is that it is a special kind of vector function, one whose curl is always zero. Since $\nabla \times \mathbf{E} = 0$, the line integral of $\mathbf{E}$ around any closed loop is zero (follows from Stokes' theorem). Because $\int \mathbf{E} \cdot d\mathbf{l} = 0$, the line integral of $\mathbf{E}$ from point $\mathbf{a}$ to point $\mathbf{b}$ is the same for all paths. Define a function

$$V(\mathbf{r}) = -\int_O^{\mathbf{r}} \mathbf{E} \cdot d\mathbf{l}. \qquad (2.25)$$

Here O is some standard reference point; V then depends only on the point $\mathbf{r}$. It is called the electric potential. It is very clear that the potential difference between two points $\mathbf{a}$ and $\mathbf{b}$ is

$$V(\mathbf{b}) - V(\mathbf{a}) = -\int_O^{\mathbf{b}} \mathbf{E} \cdot d\mathbf{l} + \int_O^{\mathbf{a}} \mathbf{E} \cdot d\mathbf{l}$$

$$= -\int_O^{\mathbf{b}} \mathbf{E} \cdot d\mathbf{l} - \int_{\mathbf{a}}^{O} \mathbf{E} \cdot d\mathbf{l} = -\int_{\mathbf{a}}^{\mathbf{b}} \mathbf{E} \cdot d\mathbf{l}. \qquad (2.26)$$

The fundamental theorem for gradients states that

$$V(\mathbf{b}) - V(\mathbf{a}) = \int_{\mathbf{a}}^{\mathbf{b}} (\nabla V) \cdot d\mathbf{l}, \qquad (2.27)$$

so that

$$\int_a^b (\nabla V) \cdot d\mathbf{l} = -\int_a^b \mathbf{E} \cdot d\mathbf{l}. \tag{2.28}$$

Since the above result is true for all the points $\mathbf{a}$ and $\mathbf{b}$, the integrands must be equal:

$$\mathbf{E} = -\nabla V. \tag{2.29}$$

Equation (2.29) is the differential version of (2.25), which says that the electric field is the gradient of a scalar potential. The original superposition principle of electrodynamics pertains to the force on a test charge Q. It says that the total force on Q is the vector sum of the forces attributable to the source charges individually:

$$\mathbf{F} = \mathbf{F}_1 + \mathbf{F}_2 + \cdots. \tag{2.30}$$

Dividing throughout by Q, we find that the electric field also obeys the superposition principle:

$$\mathbf{E} = \mathbf{E}_1 + \mathbf{E}_2 + \cdots. \tag{2.31}$$

Integrating from the common reference point to $\mathbf{r}$, it follows that the potential also satisfies such a principle:

$$V = V_1 + V_2 + \dots. \tag{2.32}$$

That is, the potential at any given point is the sum of the potentials due to all the source charges separately. This is an ordinary sum and not a vector sum.

We have shown that the electric field can be written as the gradient of a scalar potential,

$$\mathbf{E} = -\nabla V. \tag{2.33}$$

Let's see what the fundamental equations for $\mathbf{E}$

$$\nabla \cdot \mathbf{E} = \frac{\rho}{\varepsilon} \tag{2.34}$$

and

$$\nabla \times \mathbf{E} = 0 \tag{2.35}$$

look like:

$$\nabla \cdot \mathbf{E} = \nabla \cdot (-\nabla V) = -\nabla^2 V. \tag{2.36}$$

But for the negative sign, the divergence of $\mathbf{E}$ is the Laplacian of V. Gauss's law implies

$$\nabla^2 V = -\frac{\rho}{\varepsilon}. \tag{2.37}$$

Equation (2.37) is the well-known Poisson equation. In regions where there is no charge, that is, $\rho = 0$, Poisson's equation reduces to the Laplace equation,

$$\nabla^2 V = 0. \tag{2.38}$$

2.4   Ampere's Law

The electric field $\mathbf{E}$ at a point in space has been defined as the electric force $\mathbf{F_e}$ per unit charge acting on a test charge when placed at a point. In a similar fashion, we can define the magnetic flux density $\mathbf{B}$ at a point in space in terms of the magnetic force $\mathbf{F_m}$ that would be exerted on a charged particle moving with a velocity $\mathbf{u}$ were it to be passing through that point. The magnetic force $\mathbf{F_m}$ acting on a particle of charge q can be cast in the form

$$\mathbf{F_m} = q\mathbf{u} \times \mathbf{B}. \tag{2.39}$$

For a positively charged particle, the direction of $\mathbf{F_m}$ is in the direction of the cross product of $\mathbf{u} \times \mathbf{B}$, which is perpendicular to the plane containing $\mathbf{u}$ and $\mathbf{B}$ and governed by the right-hand rule. If q is negative, the direction of $\mathbf{F_m}$ is reversed. The magnitude of $\mathbf{F_m}$ is given by

$$\mathbf{F_m} = q\mathbf{uB}\sin\theta, \tag{2.40}$$

where θ is the angle between $\mathbf{u}$ and $\mathbf{B}$. It is easy to check that $\mathbf{F_m}$ is at a maximum when $\mathbf{u}$ is perpendicular to $\mathbf{B}$ ($\theta = 90°$), and it is zero when $\mathbf{u}$ is parallel to $\mathbf{B}$ ($\theta = 0$ or $180°$).

If a charged particle is in the presence of both an electric field $\mathbf{E}$ and a magnetic field $\mathbf{B}$, then the total electromagnetic force acting on it is given by

$$\mathbf{F} = \mathbf{F_e} + \mathbf{F_m} = q\mathbf{E} + q\mathbf{u} \times \mathbf{B} = q(\mathbf{E} + \mathbf{u} \times \mathbf{B}). \tag{2.41}$$

The force expressed by Eq. (2.41) is known as the Lorentz force. Electric and magnetic forces exhibit a number of important differences: (1) Whereas the electric force is always in the direction of the electric field, the magnetic force is always perpendicular to the magnetic field; (2) whereas the electric force acts on a charged particle whether or not it is moving, the magnetic force acts on it only when it is in motion; (3) whereas the electric force expands energy in displacing a charged particle, the magnetic force does no work when a particle is displaced.

The magnetic force $\mathbf{F_m}$ is always perpendicular to $\mathbf{u}$, $\mathbf{F_m} \cdot \mathbf{u} = 0$. Hence, the work performed when a particle is displaced by a differential distance $\mathbf{dl} = \mathbf{u}dt$ is given by

$$dW = \mathbf{F_m} \cdot \mathbf{dl} = (\mathbf{F_m} \cdot \mathbf{u})dt = 0. \tag{2.42}$$

Since no work is done, a magnetic field cannot change the kinetic energy of a charged particle; the magnetic field can change the direction of motion of a charged particle, but it cannot change its speed.

Another important consequence of the magnetic field is the following: The total magnetic force on any closed current loop in a uniform magnetic field is zero; that is,

$$\mathbf{F_m} = I\left(\int_c \mathbf{dl}\right) \times \mathbf{B} = 0, \tag{2.43}$$

where I is the current placed in a uniform magnetic field $\mathbf{B}$ and $\mathbf{dl}$ is the displacement vector.

Until now, we have used the magnetic flux density **B** to denote the presence of a magnetic field in a given region of space. In what follows, we shall define the magnetic field intensity, denoted by **H** and defined as being proportional to **B**; namely,

$$\mathbf{B} = \mu\mathbf{H}, \tag{2.44}$$

where the magnetic permeability μ is assumed to be known. It was established by Hans Oersted that currents induce magnetic fields that form closed loops around the wires. Based on the results obtained by Oersted, Jean Biot and Felix Savart arrived at an expression that relates the magnetic field **H** at any point in space to the current I that generates **H**. The famous Biot–Savart law states that the differential magnetic field d**H** generated by a steady current I flowing through a differential length d**l** is given by

$$d\mathbf{H} = \frac{I}{4\pi}\frac{d\mathbf{l} \times \hat{R}}{R^2}, \tag{2.45}$$

where $\mathbf{R} = \hat{R}R$ is the distance vector between d**l** and the observation point P. It is important to remember that the direction of the magnetic field is defined such that d**l** is along the direction of the current I and the unit vector $\hat{\mathbf{R}}$ points from the current element to the observation point. According to Eq. (2.45), d**H** varies as R^{-2}, which is similar to the distance dependence of the electric field induced by an electric charge. However, unlike the electric field vector **E**, whose direction is along the distance vector **R** joining the charge to the observation point, the magnetic field **H** is orthogonal to the plane containing the direction of the current element d**l** and the distance vector **R**. In order to determine the total magnetic field **H**, due to a conductor of finite size, we need to sum up the contributions due to all current elements making up the conductor. Thus, the Biot–Savart law can be written as

$$\mathbf{H} = \frac{I}{4\pi}\int_l \frac{d\mathbf{l} \times \hat{R}}{R^2}, \tag{2.46}$$

where l is the line path along which I exists. The Biot–Savart law can also be expressed in terms of the volume current density **J** or surface current density $\mathbf{J}_s$. The surface current density $\mathbf{J}_s$ applies to currents that flow on the surfaces of conductors in the form of sheets of effectively zero thickness. When the current sources are specified in terms of $\mathbf{J}_s$ over a surface S or in terms of **J** over a volume V, we can use the equivalence given by

$$I d\mathbf{l} = \mathbf{J}_s ds = \mathbf{J} dV \tag{2.47}$$

so that the Biot–Savart law can be written as

$$\mathbf{H} = \frac{I}{4\pi}\int_S \frac{\mathbf{J}_s \times \hat{\mathbf{R}}}{R^2} ds \tag{2.48}$$

$$\mathbf{H} = \frac{I}{4\pi}\int_V \frac{\mathbf{J} \times \hat{\mathbf{R}}}{R^2} dV. \tag{2.49}$$

We have talked about Gauss's law for electricity and expressed it mathematically in differential form as

$$\nabla \cdot \mathbf{E} = \frac{1}{\varepsilon}\rho.$$

The magnetic analog to a point charge is a magnetic pole, but whereas electric charges can exist in isolation, magnetic poles do not in general. Magnetic poles always occur in pairs; no matter how many times a permanent magnetic is subdivided, each new piece will always have a north pole and a south pole, even if the process were to be continued down to the atomic level. Thus, there is no magnetic equivalence to a charge Q or a charge density ρ, and it is therefore surprising that Gauss's law for magnetism is given by

$$\nabla \cdot \mathbf{B} = 0, \tag{2.50}$$

which is the differential form. The property described by Eq. (2.50) is usually called the law of nonexistence of isolate monopoles, the law of conservation of magnetic flux, or Gauss's law for magnetism. The main difference between Gauss's law for electricity and its counterpart for magnetism may be viewed in terms of the field lines. Electric field lines originate from positive electric charges and terminate on negative electric charges. Hence, for the electric field lines of the electric dipole, the electric flux through a closed surface surrounding one of the charges is not zero. However, in contrast, magnetic field lines always form continuous closed loops. Because the magnetic field lines form closed loops, the net magnetic flux through the closed surface surrounding the south pole of the magnet (or through any other closed surface) is always zero, regardless of the shape of that surface.

Consider the relationship between the magnetic field and the current as shown below:

$$\nabla \times \mathbf{H} = \mathbf{J}. \tag{2.51}$$

The integral form of the above equation obtained by integrating over an open surface S and invoking Stokes' theorem leads to the following result:

$$\int_S (\nabla \times \mathbf{H}) \cdot d\mathbf{s} = \int_S \mathbf{J} \cdot d\mathbf{s} \tag{2.52}$$

and

$$\int_C \mathbf{H} \cdot d\mathbf{l} = I, \tag{2.53}$$

the famous Ampere's law, where C is the closed contour bounding the surface S and $I = \int \mathbf{j} \cdot d\mathbf{s}$ is the total current flowing through S($\mathbf{j}$, the current density). The sign convention for the direction of C is taken so that I and $\mathbf{H}$ satisfy the right-hand rule. That is, if the direction of I is aligned with the direction of the thumb of the right hand, then the direction of the contour C should be chosen to be along the direction of the other four fingers. In words, Ampere's circuital law states that the line integral of $\mathbf{H}$ around a closed path is equal to the current traversing the surface bounded by that path.

2.5 Faraday's Law

Michael Faraday, the famous inventor, conducted some experiments with electricity and magnetism in 1831 and found that (1) pulling a loop of wire to the right through a magnetic field resulted in the flow of a current in the loop; (2) pulling to the left resulted in the same; (3) with the loop and magnet at rest, he changed the strength of the field, which once again resulted in the passage of a current. This ingenious experiment led to the result that a change in the magnetic field induces an electric field. Writing it in mathematical terms (Griffiths 1994) gives

$$\int \mathbf{E} \cdot d\mathbf{l} = -\frac{d\Phi}{dt}. \tag{2.54}$$

$\mathbf{E}$ can be related to the change in $\mathbf{B}$ by the equation

$$\int \mathbf{E} \cdot d\mathbf{l} = -\int \frac{\partial \mathbf{B}}{\partial t} \cdot d\mathbf{a}, \tag{2.55}$$

where Φ is the flux and da is the elemental area. The above expression is Faraday's law, in integral form. Applying Stokes' theorem, we can convert the above expression into a differential form as

$$\nabla \times \mathbf{E} = -\frac{\partial \mathbf{B}}{\partial t}. \tag{2.56}$$

It should be noted that Faraday's law reduces to the case $\int \mathbf{E} \cdot d\mathbf{l} = 0$. In differential form, $\nabla \times \mathbf{E} = 0$. in the static case, constant $\mathbf{B}$.

The importance of Faraday's law is that it tells us that there are two distinct kinds of electric fields: those that attribute directly to electric charges, and those associated with a change in the magnetic fields. The former can be calculated (in the static case) using Coulomb's law; the latter can be found by exploiting the analogy between Faraday's law and Ampere's law. For example,

$$\nabla \times \mathbf{E} = -\frac{\partial \mathbf{B}}{\partial t}$$

and

$$\nabla \times \mathbf{B} = \mu \mathbf{J}.$$

From Gauss's law, we have

$$\nabla \cdot \mathbf{E} = 0,$$

and for magnetic fields,

$$\nabla \cdot \mathbf{B} = 0.$$

It is interesting to note that Faraday-induced electric fields are determined by $-\partial B/\partial t$ in exactly the same way as magnetostatic fields are determined by $\mu \mathbf{J}$.

An important concept pertaining to magnetic fields is energy, which we will discuss presently. The total work done W per unit time for the charge per unit time passing down a wire I is given by

$$\frac{dW}{dt} = LI\frac{dI}{dt},$$

(2.57)

where L is the inductance. If we start with zero current and build it up to a final value I, the work done (integrating the above equation over time) is given by

$$W = \frac{1}{2}LI^2.$$

(2.58)

The above expression clearly tells us that it does not depend on how long one takes to crank up the current. It depends mainly on the geometry of the loop and the final current I. The flux Φ through the loop can be written as

$$\Phi = \int_S \mathbf{B} \cdot \mathbf{da} = \int_S (\nabla \times \mathbf{A}) \cdot \mathbf{da} = \int_C \mathbf{A} \cdot \mathbf{dl},$$

(2.59)

where C is the perimeter of the loop and S is any surface bounded by C. Thus,

$$LI = \int_C \mathbf{A} \cdot \mathbf{dl}.$$

(2.60)

The expression for W is written as

$$W = \frac{1}{2}I \int_C \mathbf{A} \cdot \mathbf{dl},$$

(2.61)

where $\mathbf{A}$ is the vector potential and dl is the line element. The above expression may be rewritten as

$$W = \frac{1}{2} \int (\mathbf{A} \cdot \mathbf{I})dl.$$

The volume current can be generalized as

$$W = \frac{1}{2} \int_V (\mathbf{A} \cdot \mathbf{J})d\tau.$$

(2.62)

Using Ampere's law, we can eliminate $\mathbf{J}$ to obtain

$$W = \frac{1}{2\mu} \int \mathbf{A} \cdot (\nabla \times \mathbf{B})d\tau.$$

(2.63)

Using the vector identities and integration by parts leads to

$$W = \frac{1}{2\mu} \left[\int_V B^2 d\tau - \int_S (\mathbf{A} \times \mathbf{B}) \cdot \mathbf{da} \right],$$

(2.64)

where S is the surface bounding the volume V. Integrating over all space, the surface integral tends to zero, with the result that

$$W = \frac{1}{2\mu} \int_{\text{allspace}} B^2 d\tau. \tag{2.65}$$

The above energy is stored in the magnetic field in the amount $(B^2/2\mu)$ per unit volume. To summarize, the energy in both electrical and magnetic fields is written as

$$W_{\text{elec}} = \frac{\varepsilon}{2} \int E^2 d\tau \tag{2.66}$$

and

$$W_{\text{mag}} = \frac{1}{2\mu} \int B^2 d\tau. \tag{2.67}$$

2.6   Vector Magnetic Potential

Earlier in the chapter, we discussed the electrostatic potential V and defined it in terms of the line integral of the electric field $\mathbf{E}$. In differential form, V and $\mathbf{E}$ are related by $\mathbf{E} = -\nabla V$. Here we will define $\mathbf{B}$ in terms of a magnetic potential with the constraint that the divergence of $\mathbf{B}$ is always equal to zero. This can be realized by taking advantage of the vector identity, which states that for any vector $\mathbf{A}$,

$$\nabla \cdot (\nabla \times \mathbf{A}) = 0. \tag{2.68}$$

Thus, by defining the vector magnetic potential $\mathbf{A}$ such that

$$\mathbf{B} = \nabla \times \mathbf{A}, \tag{2.69}$$

we are guaranteed that $\nabla \cdot \mathbf{B} = 0$. For $\mathbf{B} = \mu \mathbf{H}$, the differential form of Ampere's law reduces to

$$\nabla \times \mathbf{B} = \mu \mathbf{J}, \tag{2.70}$$

where $\mathbf{J}$ is the current density due to free charges in motion. Substituting the expression for $\mathbf{B}$ into the above equation, we have

$$\nabla \times (\nabla \times \mathbf{A}) = \mu \mathbf{J}. \tag{2.71}$$

For any vector $\mathbf{A}$, the Laplacian of $\mathbf{A}$ obeys the following identity:

$$\nabla^2 \mathbf{A} = \nabla(\nabla \cdot \mathbf{A}) - \nabla \times (\nabla \times \mathbf{A}), \tag{2.72}$$

where, by definition, $\nabla^2 \mathbf{A}$ in Cartesian coordinates is given by

$$\nabla^2 \mathbf{A} = \left(\frac{\partial^2}{\partial x^2} + \frac{\partial^2}{\partial y^2} + \frac{\partial^2}{\partial z^2} \right) \mathbf{A}$$

$$= \hat{x}\nabla^2 A_x + \hat{y}\nabla^2 A_y + \hat{z}\nabla^2 A_z.$$

Combining Eqs. (2.71) and (2.72), we have

$$\nabla(\nabla \cdot \mathbf{A}) - \nabla^2 \mathbf{A} = \mu \mathbf{J}. \tag{2.73}$$

The only constraint on the definition of $\mathbf{A}$ is that it should satisfy the condition $\nabla \cdot \mathbf{B} = 0$. There is a term $\nabla \cdot \mathbf{A}$ in the above equation. If we set $\nabla \cdot \mathbf{A} = 0$, then the equation for the magnetic potential reduces to

$$\nabla^2 \mathbf{A} = -\mu \mathbf{J}. \tag{2.74}$$

The Poisson equation (above) can be split into three scalar Poisson equations as (Jackson 1975)

$$\nabla^2 A_x = -\mu J_x$$

$$\nabla^2 A_y = -\mu J_y$$

$$\nabla^2 A_y = -\mu J_y. \tag{2.75}$$

In electrostatics, Poisson's equation for the scalar potential V is given by

$$\nabla^2 V = -\frac{\rho_V}{\varepsilon},$$

and its solution for a volume charge distribution ρ_V occupying a volume V' is given as

$$V = \frac{1}{4\pi\varepsilon} \int_{V'} \frac{\rho_V}{R'} dV'.$$

Poisson's equations for A_x, A_y, and A_z are mathematically identical in form to Poisson's equation in electrostatics. Thus, for a current density $\mathbf{J}$ with x-component J_x distributed over a volume V', the solution is given by

$$A_x = \frac{\mu}{4\pi} \int_{V'} \frac{J_x}{R'} dV'. \tag{2.76}$$

Similar solutions can be written for A_y and A_z in terms of J_y and J_z. The three solutions can be combined into a vector equation of the form

$$\mathbf{A} = \frac{\mu}{4\pi} \int_{V'} \frac{\mathbf{J}}{R'} dV'. \tag{2.77}$$

If the current distribution is given in the form of a surface current density $\mathbf{J}_s$ over a surface S', then $\mathbf{J}dV'$ should be replaced with $\mathbf{J}_s ds'$ and V' should be replaced by S'. The vector magnetic potential provides an approach for computing the magnetic field due to current-carrying conductors.

2.7 Maxwell's Equations

Maxwell's equations are based on the basic equations of electro- and magnetostatics. For example, we know that the electric charges are the sources and sinks of the vector field of the dielectric displacement density D. Thus, the flux of the dielectric displacement through a surface enclosing the charge is given by

$$\frac{1}{4\pi} \int_{\text{area}} \mathbf{D} \cdot \mathbf{n} da = \int_V \rho dV, \tag{2.78}$$

with $\mathbf{n}$ the unit normal, which is essentially a simplification of Coulomb's force law. From Faraday's induction law, we have

$$V = \int \mathbf{E} \cdot d\mathbf{r} = -\frac{1}{c} \frac{\partial \phi}{\partial t}. \tag{2.79}$$

ϕ in the above equation is defined as $\phi = \int_{\text{area}} \mathbf{B} \cdot \mathbf{n} da$. The absence of isolated monopoles implies that

$$\int_{\text{area}} \mathbf{B} \cdot \mathbf{n} da = 0. \tag{2.80}$$

The implications are that the magnetic induction is source-free and that the field lines are closed curves. As already stated, Ampere's law, written in terms of a mathematical expression, has the form

$$\int \mathbf{H} \cdot d\mathbf{r} = \frac{4\pi}{c} \int \mathbf{J} \cdot \mathbf{n} da. \tag{2.81}$$

Maxwell's equation in integral representation can be written immediately as

$$\int_{\text{area}} \mathbf{D} \cdot \mathbf{n} da = 4\pi \int_V \rho dV$$

$$\int \mathbf{E} \cdot d\mathbf{r} = -\frac{1}{c} \frac{\partial}{\partial t} \int \mathbf{B} \cdot \mathbf{n} da$$

$$\int_{\text{area}} \mathbf{B} \cdot \mathbf{n} da = 0$$

$$\int \mathbf{H} \cdot d\mathbf{r} = \frac{4\pi}{c} \left(\int_{\text{area}} \mathbf{J} \cdot \mathbf{n} da + \frac{1}{4\pi} \frac{d}{dt} \int_{\text{area}} \mathbf{D} \cdot \mathbf{n} da \right).$$

Table 2.1 Maxwell's equations

Reference	Differential form	Integral form
Gauss's law	$\nabla \cdot \mathbf{D} = 4\pi\rho$	$\int_S \mathbf{D} \cdot s = Q$
Faraday's law	$\nabla \times \mathbf{E} = -(1/c)\partial \mathbf{B}/\partial t$	$\int_C \mathbf{E} \cdot \mathbf{dl} = -\int_S \partial \mathbf{B}/\partial t \cdot \mathbf{ds}$
No magnetic charges	$\nabla \cdot \mathbf{B} = 0$	$\int_S \mathbf{B} \cdot \mathbf{ds} = 0$
Ampere's law	$\nabla \times \mathbf{H} = (4\pi/c)$	$\int_C \mathbf{H} \cdot \mathbf{dl} = \int_S (4\pi \mathbf{J}/c$
	$(\mathbf{J} + (1/4\pi)\partial \mathbf{D}/\partial t)$	$+ (1/c)\partial \mathbf{D}/\partial t) \cdot \mathbf{ds}$

Applying Gauss's and Stokes' theorems, we can simplify the above expressions to yield them in differential form as follows:

$$\nabla \cdot \mathbf{D} = 4\pi\rho \tag{2.82}$$

$$\nabla \times \mathbf{E} = -\frac{1}{c}\frac{\partial \mathbf{B}}{\partial t} \tag{2.83}$$

$$\nabla \cdot \mathbf{B} = 0 \tag{2.84}$$

$$\nabla \times \mathbf{H} = \frac{4\pi}{c}\left(\mathbf{J} + \frac{1}{4\pi}\frac{\partial \mathbf{D}}{\partial t}\right). \tag{2.85}$$

In addition to the above equations, one has to consider the continuity equation (conservation of mass) and the force law equation given below:

$$\nabla \cdot \mathbf{J} + \frac{\partial \rho}{\partial t} = 0 \tag{2.86}$$

$$\mathbf{f} = \rho \mathbf{E} + \frac{1}{c}\mathbf{J} \times \mathbf{B}. \tag{2.87}$$

Maxwell's equations are partial, linear, coupled differential equations of the first order. Due to linearity, the principle of superposition is valid.

The Maxwell equations are made up of a set of coupled first-order partial differential equations relating the various electric and magnetic fields (Table 2.1). It is convenient to introduce potentials, obtaining a smaller number of second-order equations, which satisfy the Maxwell equations identically. Let's introduce the scalar potential Φ and vector potentiall $\mathbf{A}$ and see what the equations reduce to. Since $\nabla \cdot \mathbf{B} = 0$ holds, $\mathbf{B}$ can be defined in terms of a vector potential:

$$\mathbf{B} = \nabla \times \mathbf{A}. \tag{2.88}$$

The homogeneous Faraday law can be written as

$$\nabla \times \left(\mathbf{E} + \frac{1}{c}\frac{\partial \mathbf{A}}{\partial t}\right) = 0. \tag{2.89}$$

The above expression implies that the terms with a vanishing curl can be written as the gradient of some scalar function, namely, a scalar potential Φ as

$$\mathbf{E} + \frac{1}{c}\frac{\partial \mathbf{A}}{\partial t} = -\nabla\Phi$$

or

$$\mathbf{E} = -\nabla\Phi - \frac{1}{c}\frac{\partial \mathbf{A}}{\partial t}. \tag{2.90}$$

The expressions for $\mathbf{B}$ and $\mathbf{E}$ in terms of the potentials $\mathbf{A}$ and Φ identically satisfy the two homogeneous Maxwell equations. The dynamic behavior of $\mathbf{A}$ and Φ will be determined by the two inhomogeneous equations, which can be written in terms of the potentials as

$$\nabla^2\Phi + \frac{1}{c}\frac{\partial}{\partial t}(\nabla \cdot \mathbf{A}) = -4\pi\rho \tag{2.91}$$

$$\nabla^2\mathbf{A} - \frac{1}{c^2}\frac{\partial^2 \mathbf{A}}{\partial t^2} - \nabla\left(\nabla \cdot \mathbf{A} + \frac{1}{c}\frac{\partial \Phi}{\partial t}\right) = -\frac{4\pi}{c}\mathbf{J}. \tag{2.92}$$

The four Maxwell equations have now been reduced to two equations. However, they are coupled equations. The uncoupling needs to be addressed by exploiting the arbitrariness involved in the definition of the potentials. Introduce the transformation

$$\mathbf{A} \rightarrow \mathbf{A}' = \mathbf{A} + \nabla\Lambda \tag{2.93}$$

and the following for the scalar potential:

$$\Phi \rightarrow \Phi' = \Phi - \frac{1}{c}\frac{\partial \Lambda}{\partial t}. \tag{2.94}$$

The expressions (2.93) and (2.94) give one a chance to choose $(\mathbf{A}, \Phi)$ such that

$$\nabla \cdot \mathbf{A} + \frac{1}{c}\frac{\partial \Phi}{\partial t} = 0. \tag{2.95}$$

The two equations for Φ and $\mathbf{A}$ will get decoupled, leaving two inhomogeneous wave equations for Φ and $\mathbf{A}$, as follows:

$$\nabla^2\Phi - \frac{1}{c^2}\frac{\partial^2 \Phi}{\partial t^2} = -4\pi\rho \tag{2.96}$$

and

$$\nabla^2\mathbf{A} - \frac{1}{c^2}\frac{\partial^2 \mathbf{A}}{\partial t^2} = -\frac{4\pi}{c}\mathbf{J}. \tag{2.97}$$

Equations (2.95)–(2.97) form a set of equations equivalent in all respects to Maxwell's equations.

Chapter 3
MHD Equations and Concepts

3.1 Assumptions

In Chap. 2, we discussed some of the basic concepts of electricity and magnetism in general, the equations governing them, and the results connecting them. However, there was no specific mention of application to the Sun. Also, the discussion was mostly on statics rather than dynamics. We all know that the Sun is a continuously evolving, dynamic hot plasma in which the magnetic fields play a very key role. In this chapter, we will discuss the definition and properties of magnetic flux tubes, diffusion of magnetic fields, and some simple analytic solutions of magnetohydrodynamic (MHD) equations relevant to the Sun. A brief discussion on the Parker solution of solar wind will also be presented.

To begin with, let's define the basic parameters pertaining to plasma and move on to describe the MHD equations, the assumptions involved, and some justification for employing these equations to the Sun. More information on the basics of MHD may be found in Schnack (2009). Most of the structures observed and interesting phenomena taking place on the Sun are due to the magnetic field prevailing everywhere. In particular, the solar atmosphere is far from being static and uniform; it is highly complex with an inhomogeneous environment, and there are many examples of the solar plasma interacting with the magnetic field.

There are three parameters that characterize a plasma, namely, the particle density, the temperature, and the magnetic field. Some of the assumptions that are made in studying plasma are the following: 1. The plasma is a continuum, a situation when the typical length scale of the system exceeds the ion gyroradius. This is certainly valid for the different phenomena that we discuss in this book. 2. The plasma is a single fluid, which is true if the length scales of the system are much longer than the Debye shielding length. 3. The plasma is in thermodynamic equilibrium with a distribution closer to a Maxwellian. This is possible if the time scale of the system is larger than the typical collision time scale and if the length

A. Satya Narayanan, *An Introduction to Waves and Oscillations in the Sun*, Astronomy and Astrophysics Library, DOI 10.1007/978-1-4614-4400-8_3,

scales considered are more than the mean free path. 4. Relativistic effects may be neglected. 5. Changes in the permeability, conductivity, or thermal diffusivity are not significant, that is, they are more isotropic.

A plasma in general can be modeled by three descriptions—Vlasov, two-fluid, and MHD. Of the three descriptions mentioned here, the Vlasov description is the most accurate, while the MHD is the least accurate. (Interested readers may look into the book by Dendy (1990) for more details.) This being the case, why is it that one resorts to MHD? The reason is simple: MHD is more of a macroscopic point of view, and in situations where greater detail and accuracy are not required, it is appropriate to use the MHD description. It is easily amenable to model complex geometries. For example, the equilibrium and stability of three-dimensional flows of a finite extent are described using MHD models. The finer points using the Vlasov and two-fluid models can be worked out once we have an approximate understanding of the physical processes taking place in a system using the MHD description. The MHD models are more useful and appropriate when one deals with a system where the magnetic field plays a dominant role. In particular, the Sun, in which the magnetic field plays a major role in evolution, can be better understood using the MHD approach. Some areas where MHD is used widely are the solar and astrophysical plasmas, planetary and stellar dynamos, magnetospheric physics of planets, and stars (pulsars in particular).

The MHD theory, which is a single-particle picture, can be described by a theory that provides information on the dynamics of a group of particles. The theory dealing with the dynamics of a group of particles is the classic theory of fluids. Under certain restrictions, a collection of charged particles can be treated as a fluid, in particular, an MHD fluid. This deals with the motion of particles in the presence of electromagnetic fields. It ignores the identity of individual particles and considers the fluid element. The motion of an ensemble of these particles constitutes a fluid motion. An important element of the MHD theory is that it incorporates the effects that arise from the motion of an electrically conducting fluid across magnetic fields. The collective interaction involving motion, currents, and magnetic fields characterizes the general behavior of MHD fields. The theory of MHD was developed to describe the observation and discovery of sunspots in the early nineteenth century. It provided a strong forum for the generation and maintenance of magnetic fields on the Sun. An excellent introduction to MHD is found in Goedbloed and Poedts (2004). A more recent book on MHD is Goedbloed et al. (2010).

One of the most important characteristics of MHD is the units involved for the different quantities appearing in describing the fluid and the external forces that act on them. From what follows, we shall briefly describe some of the units that appear in the MHD description:

The length is measured in terms of meters (1 m) or kilometers, depending on the situation describing the physical model, whereas the mass is usually expressed in terms of kilograms (1 kg). The standard representation for time is in terms of seconds (s), unlike light-years in astronomy. The force acting on a surface is described by newtons (1 N), which is $1 \, \mathrm{kg \, m \, s^{-2}}$. The current is denoted by amperes [1 amp (A)]. The magnetic induction and charge are respectively expressed as teslas (1 tesla $= 10^4$ G) and Coulombs (C). In most of the problems pertaining to MHD, the effects of electric field and displacement may be neglected.

3.2 Dimensionless Parameters

Dimensional analysis is a tool to understand the properties of physical quantities independent of the units used to measure them. Every physical quantity is some combination of mass, length, time, and electric charge (denoted M, L, T, and Q, respectively). For example, speed, which may be measured in meters per second (m/s) or miles per hour (mi/h), has the dimension L/T or, alternatively, LT^{-1}.

Dimensional analysis is routinely used to check the plausibility of derived equations and computations. It is also used to form reasonable hypotheses about complex physical situations that can be tested by experiment or by more developed theories of the phenomena. It categorizes different types of physical quantities and units based on their relationships to or dependence on other units or their dimensions, if any.

The basic principle of dimensional analysis was known as early as the days of Isaac Newton, who referred to it as the "Great Principle of Similitude." The nineteenth-century French mathematician Joseph Fourier made important contributions based on the idea that physical laws like $F = ma$ should be independent of the units employed to measure the physical variables. A dimensional equation can have the dimensions reduced or eliminated through nondimensionalization, which begins with dimensional analysis and involves scaling quantities by characteristic units of a system or natural units of nature. This gives insight into the fundamental properties of the system. The scaling laws in hydrodynamics and MHD is found in Schnack (2009).

In hydrodynamics, the most important and often used nondimensional number is the Reynolds number, which gives the ratio of the size of the inertial term to the viscous term in the equations of motion, defined by

$$\mathrm{Re} = \frac{lV}{\nu}. \tag{3.1}$$

Here V is the typical plasma speed and l the length scale. ν is the kinematic viscosity. In MHD, one can define the magnetic Reynolds number, which is a measure of the strength of the coupling between the flow and the magnetic field, as

$$R_{\mathrm{m}} = \frac{lV}{\eta}, \tag{3.2}$$

where η is the magnetic permeability. When the coupling is weak, one finds that $R_{\mathrm{m}} \ll 1$, whereas in the solar atmosphere it is rather large, $R_{\mathrm{m}} \gg 1$, which implies that the coupling is very strong.

The Mach number, which is used frequently in aerodynamics to distinguish subsonic, supersonic, and hypersonic flows, is defined as the ratio of the flow speed (V) to the sound speed c_s as

$$M = \frac{V}{c_s}, \tag{3.3}$$

where the sound speed is defined as $c_s = (\gamma p_0/\rho_0)^{1/2}$. In a similar way, the Alfvén Mach number may be defined as

$$M_A = \frac{V}{V_A}. \tag{3.4}$$

One of the important parameters that play an important role in plasma physics is the plasma beta (β), which is defined as

$$\beta = \frac{2\mu p}{B^2}. \tag{3.5}$$

The plasma beta is much larger than 1 inside the Sun, while it is of the order of 1 at the photospheric and chromospheric levels. It is much smaller than 1 at coronal heights. The plasma beta is basically a measure of the relative importance of the gas pressure to the magnetic pressure.

There are other nondimensional parameters, such as the Rossby number, Prandtl number, Rayleigh number, Chandrasekhar number, and Hartman number, to name a few. The discussion of these numbers is beyond the scope of this book and will not be dealt with here.

3.3   Mass Continuity

The behavior of a system made up of charged particles is considerably different from that of ordinary fluids. However, certain concepts and equations that govern the equations of motion of ordinary fluids are general in that they may be applicable to systems having certain charged particles by approximating them as conducting fluids.

We mentioned at the beginning of this chapter that the MHD fluid medium will be approximated as an ordinary fluid medium and so it can be treated as a continuous medium. Fluid dynamics, which deals with macroscopic phenomena, assumes that any small-volume element contains many particles. Thus, one can define the macroscopic parameters, such as the density ρ, in describing the fluid property.

Consider an arbitrary region of space consisting of an MHD fluid that occupies a volume V bounded by a surface S. Let dV be an element of this volume and dS an element of the surface. Let $\mathbf{U}(x,y,z,t)$ be the velocity of a fluid element at a given position in space (x,y,z) and time t. A fluid is displaced in time dt at a distance $\mathbf{U}dt$, and the mass of fluid crossing dS per unit time is

$$dm = \rho \mathbf{U} \cdot \mathbf{n} dS dt. \tag{3.6}$$

The total mass of the fluid flowing out of the volume V per unit time is

$$m = \int_S \rho \mathbf{U} \cdot \mathbf{n} dS, \tag{3.7}$$

where the integral extends over the surface S. This outward flow will result in a decrease of fluid contained in V, and the total amount that is diminished per unit time is

$$m = -\int_V \frac{\partial \rho}{\partial t} dV, \tag{3.8}$$

where the integral extends over the volume and the negative sign indicates that the fluid is being lost. Since there are no sources or sinks for the fluid, the above two equations can be equated to yield

$$\int_S \rho \mathbf{U} \cdot \mathbf{n} dS = -\int_V \frac{\partial \rho}{\partial t}. \tag{3.9}$$

Using the Gauss divergence theorem, one can convert the surface integral on the left-hand side of the above equation to yield

$$\int_V \left(\nabla \cdot \rho \mathbf{U} + \frac{\partial \rho}{\partial t} \right) dV = 0. \tag{3.10}$$

The integrand is continuous, and since the above relation holds for any arbitrary volume V, the integrand can be equated to zero identically; that is,

$$\frac{\partial \rho}{\partial t} + \nabla \cdot \rho \mathbf{U} = 0. \tag{3.11}$$

The above equation in the literature is known as the equation of continuity and is fundamental in hydrodynamics. The implication of the above equation is that matter is conserved, and it is valid for all fluids, irrespective of whether the fluid is adiabatic, compressional, isothermal, viscous, or turbulent. The vector $\rho \mathbf{U}$ is in the direction of the flow and represents the mass flux density. Matter should be conserved in MHD, and thus the continuity equation should be satisfied by an MHD fluid.

An important concept in hydrodynamics is the notion of convective derivative, which also holds true in MHD. The density of a fluid $\rho(x,y,z,t)$ depends on the time t explicitly and on the coordinates (x,y,z) implicitly, since the coordinate of the fluid changes with time with the displacement of particles. Thus, one realizes that the total time rate of change of the density should be

$$\frac{d\rho}{dt} = \frac{\partial \rho}{\partial t} + \frac{\partial \rho}{\partial x}\frac{dx}{dt} + \frac{\partial \rho}{\partial y}\frac{dy}{dt} + \frac{\partial \rho}{\partial z}\frac{dz}{dt}$$

$$= \frac{\partial \rho}{\partial t} + (\mathbf{U} \cdot \nabla)\rho. \tag{3.12}$$

Assuming $\mathbf{U} = \mathbf{i}dx/dt + \mathbf{j}dy/dt + \mathbf{k}dz/dt$ and $\nabla = \mathbf{i}\partial/\partial x + \mathbf{j}\partial/\partial y + \mathbf{k}\partial/\partial z$, the second line in the above equation follows. Here, $(\mathbf{i}, \mathbf{k}, \mathbf{j})$ are the unit vectors in the

Cartesian coordinate system. The total derivative in Eq. (3.12) in hydrodynamics is called the convective derivative. This relationship between the rate of change of a variable in a moving and fixed frame of reference is a general form and may be applied to any variable of the fluid, such as the velocity or magnetic field, as in MHD fluids.

A flow is classified as irrotational if $\nabla \times \mathbf{U} = 0$. For such flows, one can define a scalar function Φ such that

$$\mathbf{U} = -\nabla \Phi. \tag{3.13}$$

Φ is called a scalar potential of the flow vector $\mathbf{U}$ (irrotational flows are also called potential flows in hydrodynamics). The components of $\mathbf{U} = -\nabla \Phi$ in Cartesian coordinates are

$$\frac{\partial \Phi}{\partial x} = U_x$$

$$\frac{\partial \Phi}{\partial y} = U_y$$

$$\frac{\partial \Phi}{\partial z} = U_z \tag{3.14}$$

and the velocity potential Φ can be determined from

$$\Phi(x,y,z) = \int (U_x dx + U_y dy + U_z dz). \tag{3.15}$$

The surfaces on which Φ is a constant are called equipotential surfaces. The curves along which the potential is constant are obtained by setting

$$d\Phi(x,y,z) = 0. \tag{3.16}$$

It is interesting to note that the equipotential surfaces given by $\mathbf{U} = -\nabla \Phi$ are orthogonal to the surface of the velocity fields.

A flow in hydrodynamics is said to be incompressible if the divergence of the flow vector $\mathbf{U}$ vanishes. $\nabla \cdot \mathbf{U} = 0$. In this case, the flow vector is said to be solenoidal. Incompressible flow fields $\mathbf{U}$ can be described by a vector $\mathbf{w}$ such that

$$\mathbf{U} = \nabla \times \mathbf{w} \tag{3.17}$$

since the divergence of the curl of any vector vanishes identically. Thus, irrotational and solenoidal vectors are interesting from a mathematical point of view, because any vector field $\mathbf{A}$ can be written as $\mathbf{A} = \nabla \Phi + \nabla \times \mathbf{w}$. It can be shown by simple algebra that if a flow is both irrotational and solenoidal, then the flow potential satisfies Laplace's equation:

$$0 = \nabla \cdot \mathbf{U}$$

$$= -\nabla \cdot \nabla \Phi$$

$$= -\nabla^2 \Phi. \tag{3.18}$$

3.4 Equations of Motion

Assume the densities of the electrons and ions to be ρ_e and ρ_i, respectively. In the previous section, we showed that the plasma satisfies the equation for the conservation of mass, so that the two-component plasma satisfies the following equations:

$$\frac{\partial \rho_i}{\partial t} + \nabla \cdot \rho_i \mathbf{u}_i = 0$$

$$\frac{\partial \rho_e}{\partial t} + \nabla \cdot \rho_e \mathbf{u}_e = 0. \tag{3.19}$$

Adding the above equations, one obtains

$$\frac{\partial (\rho_i + \rho_e)}{\partial t} + \nabla \cdot (\rho_i \mathbf{u}_i + \rho_e \mathbf{u}_e) = 0. \tag{3.20}$$

Define the mass density of the fluid as

$$\rho_m = n_i m_i + n_e m_e$$

and the fluid velocity as

$$\mathbf{U} = \frac{n_i m_i \mathbf{u}_i + n_e m_e \mathbf{u}_e}{n_i m_i + n_e m_e}.$$

The above two equations can be combined to yield

$$\frac{\partial \rho_m}{\partial t} + \nabla \cdot \rho_m \mathbf{U} = 0. \tag{3.21}$$

The equation of motion for a single particle is given by

$$m \frac{d\mathbf{v}}{dt} = q(\mathbf{E} + \mathbf{v} \times \mathbf{B}). \tag{3.22}$$

We also assume that there are no thermal motions and collisions are completely ignored. This implies that all particles move together, and the fluid equation of motion of the particles is obtained by multiplying the above equation by n to give

$$mn \frac{d\mathbf{u}}{dt} = qn(\mathbf{E} + \mathbf{u} \times \mathbf{B}). \tag{3.23}$$

The velocity of the individual particle $\mathbf{v}$ is replaced by $\mathbf{u}$, which is the fluid velocity, expressed as the average velocity. Introducing the pressure forces leads to

$$mn \frac{d\mathbf{u}}{dt} = qn(\mathbf{E} + \mathbf{u} \times \mathbf{B}) - \nabla p. \tag{3.24}$$

The minus sign is introduced to emphasize the fact that the flow is driven in the direction opposite that of the pressure gradient. It also implies that the pressure is a scalar quantity and is isotropic.

One can write equations of motion for the electrons and ions by replacing the velocity and density by the subscripts "e" and "i," respectively. Define the current density $\mathbf{J}$ as

$$\mathbf{J} = n_i q_i \mathbf{u}_i + n_e q_e \mathbf{u}_e.$$

Combining the definition of the density, velocity, and current density, the equation of motion can be simplified to yield

$$\rho_m \frac{d\mathbf{U}}{dt} = \mathbf{J} \times \mathbf{B} - \nabla p. \tag{3.25}$$

It is interesting to note that the electric field does not appear explicitly in the one-fluid momentum equation. The additional term $(\mathbf{J} \times \mathbf{B})$ arises from the coupling of the current density to the magnetic field $\mathbf{B}$. It is this $\mathbf{J} \times \mathbf{B}$ force that makes the electromagnetic fluid different from the ordinary gas fluid consisting only of neutral particles. This is referred to as the electromagnetic stress tensor.

We already discussed in the previous chapter the importance of a relation between the current density, electric field, fluid motion, and magnetic field. Written explicitly, this will look like

$$\mathbf{J} = \sigma(\mathbf{E} + \mathbf{U} \times \mathbf{B}). \tag{3.26}$$

The simplified Maxwell's equation relating the electric and magnetic fields can be written as

$$\nabla \times \mathbf{E} = -\frac{\partial \mathbf{B}}{\partial t} \tag{3.27}$$

$$\nabla \times \mathbf{B} = \mu_0 \mathbf{J}. \tag{3.28}$$

μ_0 is the permeability in free space. Substituting for $\mathbf{E}$ from Eq. (3.26) in Eq. (3.27) leads to

$$\frac{\partial \mathbf{B}}{\partial t} = \nabla \times \left(\mathbf{U} \times \mathbf{B} - \frac{\mathbf{J}}{\sigma} \right). \tag{3.29}$$

Substituting for $\mathbf{J}$ from Eq. (3.28) in Eq. (3.29) and using the relationship $\lambda = 1/(\mu_0 \sigma)$ (where λ is the magnetic diffusivity), we have

$$\frac{\partial \mathbf{B}}{\partial t} = \nabla \times (\mathbf{U} \times \mathbf{B} - \lambda \nabla \times \mathbf{B}). \tag{3.30}$$

If the magnetic diffusivity is constant in space, then the above equation reduces to

$$\frac{\partial \mathbf{B}}{\partial t} = \nabla \times (\mathbf{U} \times \mathbf{B}) + \lambda \nabla^2 \mathbf{B}. \tag{3.31}$$

The above equation is one of the fundamental equations of MHD and is referred to as the induction equation. It describes the behavior of magnetic fields in a plasma system.

3.5 Energy Equation

In order to derive the energy equation, let's start with Eq. (3.25), which can be written as

$$\rho_m \frac{dU}{dt} = -\nabla p + \frac{1}{\mu_0}(\nabla \times B) \times B. \tag{3.32}$$

Taking the dot product of the above equation with U results in

$$\rho_m U \cdot \frac{dU}{dt} = -U \cdot \nabla p + \frac{U}{\mu_0} \cdot (\nabla \times B) \times B. \tag{3.33}$$

The term on the left-hand side of the above equation can be written as

$$\rho_m U \cdot \frac{dU}{dt} = \rho_m U \cdot \left(\frac{\partial}{\partial t} + U \cdot \nabla\right) U$$

$$= \frac{\rho_m}{2} \frac{\partial U^2}{\partial t} + \frac{\rho_m}{2} U \cdot \nabla U^2$$

$$= \frac{\partial}{\partial t} \frac{\rho_m U^2}{2} - \frac{U^2}{2} \frac{\partial \rho_m}{\partial t} + \frac{\rho_m}{2} U \cdot \nabla U^2. \tag{3.34}$$

Using the continuity equation and combining the second and third terms yield

$$\rho_m U \cdot \frac{dU}{dt} = \frac{\partial}{\partial t} \frac{\rho_m U^2}{2} + \nabla \cdot \frac{\rho_m U^2}{2} U. \tag{3.35}$$

If we assume that the MHD fluid is adiabatic, then the relation between the pressure and the density can be written as

$$\frac{d}{dt}(p\rho_m^{-\gamma}) = 0. \tag{3.36}$$

The first term on the right-hand side of Eq. (3.33), using the adiabatic nature of the fluid, can be simplified to yield

$$\frac{dp}{dt} - \frac{\gamma p}{\rho_m} \frac{d\rho_m}{dt} = 0. \tag{3.37}$$

The term on the left-hand side of the above equation is the convective derivative of the pressure term, which on expansion would look like

$$\frac{dp}{dt} = \frac{\partial p}{\partial t} + (U \cdot \nabla)p.$$

Using the continuity equation and the above expression, Eq. (3.37) can be simplified as

$$(1 - \gamma)(U \cdot \nabla)p + \frac{\partial p}{\partial t} + \gamma \nabla \cdot (pU) = 0. \tag{3.38}$$

The second term on the right-hand side of Eq. (3.33) can be simplified as follows:

$$\frac{1}{\mu_0}\mathbf{U}\cdot(\nabla\times\mathbf{B})\times\mathbf{B} = -\frac{1}{\mu_0}(\mathbf{U}\times\mathbf{B})\cdot(\nabla\times\mathbf{B})$$

$$= \frac{1}{\mu_0}\mathbf{E}\cdot(\nabla\times\mathbf{B}). \tag{3.39}$$

In the above equation, we have assumed that $\mathbf{E} = -\mathbf{U}\times\mathbf{B}$. For a large value of σ, the term on the right-hand side of the above equation will be very small. However, in the limit $\sigma\to\infty$, $\mathbf{J}\cdot\mathbf{E} = 0$. Using the vector relation

$$\nabla\cdot(\mathbf{E}\times\mathbf{B}) = \mathbf{B}\cdot(\nabla\times\mathbf{E} - \mathbf{E}\cdot(\nabla\times\mathbf{B})$$

in Eq. (3.39), we have the term on the right-hand side as

$$= -\frac{1}{2\mu_0}\frac{\partial B^2}{\partial t} - \frac{1}{\mu_0}\nabla\cdot(\mathbf{E}\times\mathbf{B}). \tag{3.40}$$

Using the relations (3.35), (3.38). and (3.40) in Eq. (3.33) yields the conservation of energy condition for the MHD fluid as

$$\frac{\partial}{\partial t}\left(\frac{\rho_m U^2}{2} + \frac{p}{\gamma-1} + \frac{B^2}{2\mu_0}\right) + \nabla\cdot\left(\frac{\rho_m U^2}{2}\mathbf{U} + \frac{\gamma}{\gamma-1}p\mathbf{U} + \frac{\mathbf{E}\times\mathbf{B}}{\mu_0}\right). \tag{3.41}$$

The terms in the first bracket represent the kinetic energy of the fluid motion, the thermal energy, and the total energy density of the magnetic field. The terms in the second bracket represent the rates at which these various energies are flowing.

In summary, the basic equations describing the MHD fluid are given below:

$$\nabla\times\mathbf{E} = -\frac{\partial\mathbf{B}}{\partial t}$$

$$\nabla\cdot\mathbf{B} = 0$$

$$\nabla\times\mathbf{H} = \mathbf{J}$$

$$\nabla\cdot\mathbf{D} = 0$$

$$\frac{\partial\rho_m}{\partial t} + \nabla\cdot\rho_m\mathbf{U} = 0$$

$$\rho_m\frac{d}{dt}\mathbf{U} = \mathbf{J}\times\mathbf{B} - \nabla p$$

$$\mathbf{J} = \sigma(\mathbf{E} + \mathbf{U}\times\mathbf{B})$$

$$\frac{d}{dt}(p\rho_m^{-\gamma}) = 0.$$

There are two more constitutive relations among the field variables, namely,

$$\mathbf{B} = \mu_0 \mathbf{H}$$

$$\mathbf{D} = \varepsilon_0 \mathbf{E}.$$

It is interesting to note that the induction equation is very similar to the hydrodynamic equation that describes the behavior of vorticity in an incompressible fluid, namely,

$$\frac{\partial \omega}{\partial t} = \nu \nabla^2 \omega + \nabla \times (\mathbf{U} \times \omega). \tag{3.42}$$

ω is the vorticity ($\omega = \nabla \times \mathbf{U}$) and ν is the kinematic viscosity. The first term on the right-hand side of the above equation describes the effects of diffusion and the second term the convection of the vorticity. By analogy, this interpretation may be applied to the MHD fluid. Thus, $1/\mu_0 \sigma$ can be defined as the magnetic viscosity. It is a measure of how fast the magnetic field diffuses out of (or into) the fluid, given the conductivity of the fluid.

A note of caution is that although ω and $\mathbf{B}$ are described by identical equations, one cannot infer that $\mathbf{B}$ is completely analogous to ω in hydrodynamics. In hydrodynamics, the vorticity and velocity are related, while no such relation exists for the magnetic field.

In what follows, let's discuss the behavior of the magnetic field for two special cases, namely, the diffusion of the magnetic field and the ideal MHD fluid wherein the conductivity σ is assumed to be infinite. The starting point for the evolution of the magnetic field would be the induction equation as described by Eq. (3.31).

For the first case, let's set $\mathbf{U} = 0$ in Eq. (3.31). The time variation of the magnetic field is then described by

$$\frac{\partial \mathbf{B}}{\partial t} = \lambda \nabla^2 \mathbf{B}, \tag{3.43}$$

where λ has been defined earlier. This equation looks similar to the heat conduction equation

$$\frac{\partial T}{\partial t} = \kappa \nabla^2 T, \tag{3.44}$$

where κ is the coefficient of thermal conduction, and with the vorticity equation,

$$\frac{\partial \omega}{\partial t} = \nu \nabla^2 \omega. \tag{3.45}$$

These equations, which show how $\mathbf{B}$, T, and ω change in time relative to spatial changes, describe the diffusion of the magnetic field, heat, and vorticity.

The solution of the induction equation for the first case, assuming that each Cartesian component of the magnetic fieldd diffuses with time from its initial configuration $B_i(\mathbf{r}, 0)$, can be obtained from the Green function as

$$G(\mathbf{r} - \mathbf{r}', t) = (4\pi \lambda t)^{-3/2} \exp\left[-\frac{(\mathbf{r} - \mathbf{r}')^2}{4\lambda t}\right] \tag{3.46}$$

and the magnetic field for $t > 0$ is obtained from the integral equation

$$B_i(\mathbf{r},t) = \int \int \int G(\mathbf{r} - \mathbf{r}',t)B_i(\mathbf{r}',0)\mathrm{d}^3\mathbf{r}'. \tag{3.47}$$

An estimate of how rapidly or slowly the diffusion is occurring can be obtained by assuming L to be the characteristic spatial length scale of $\mathbf{B}$. Then one can substitute L^{-2} for ∇^2 to obtain

$$\frac{\partial \mathbf{B}}{\partial t} \approx \pm \frac{1}{\mu_0 \sigma L^2} \mathbf{B}, \tag{3.48}$$

where the $\pm$ sign refers to the gain or loss of the magnetic field with time. The solution of the simplified equation is written as

$$\mathbf{B} = \mathbf{B}_0 e^{\pm t/t_d}, \tag{3.49}$$

where $\mathbf{B}_0$ is the initial value of the magnetic field and

$$t_D = \mu \sigma L^2 \tag{3.50}$$

is the characteristic time for the magnetic field to increase or decay to $1/e$ of its initial value.

Another simple solution for the diffusion of the magnetic field for the induction equation is the following: Consider the diffusion of a unidirectional magnetic field $\mathbf{B} = B(x,t)e_y$ with the initial step-function profile (Nakariakov 2002)

$$B(x,0) = +B_0 \quad x > 0$$
$$B(x,0) = -B_0 \quad x < 0, \tag{3.51}$$

which is like a current sheet. For a unidirectional magnetic field, the induction equation reduces to

$$\frac{\partial B}{\partial t} = \lambda \frac{\partial^2 B}{\partial x^2}. \tag{3.52}$$

In addition to the initial conditions, the following boundary condition needs to be incorporated:

$$B(\pm\infty,t) = \pm B_0. \tag{3.53}$$

The solution of the above equation satisfying the boundary conditions is given by

$$B(x,t) = B_0 erf(\xi), \tag{3.54}$$

where $\xi = x/(4\lambda t)^{1/2}$ and

$$erf(\xi) = \frac{2}{\pi^{1/2}} \int_0^\xi \exp(-u^2)\mathrm{d}u$$

is the error function.

For the second case of an ideal MHD fluid where the conductivity σ is infinite, the magnetic field satisfies the following equation:

$$\frac{\partial \mathbf{B}}{\partial t} = \nabla \times (\mathbf{U} \times \mathbf{B}). \tag{3.55}$$

Comparing this equation with Maxwell's equation $\partial \mathbf{B}/\partial t = -\nabla \times \mathbf{E}$, one can see that the electric field in an infinitely conducting medium is given by

$$\mathbf{E} = -\mathbf{U} \times \mathbf{B}. \tag{3.56}$$

Write $\mathbf{E} = \mathbf{E}_{\parallel} + \mathbf{E}_{\perp}$ and $\mathbf{U} = \mathbf{U}_{\parallel} + \mathbf{U}_{\perp}$, where $\parallel$ and $\perp$ refer to the directions relative to $\mathbf{B}$. Inserting the above expression into Eq. (3.56), we have

$$\mathbf{E}_{\parallel} = -\mathbf{U}_{\parallel} \times \mathbf{B}$$
$$= 0$$
$$\mathbf{E}_{\perp} = -\mathbf{U}_{\perp} \times \mathbf{B}. \tag{3.57}$$

The relation (3.56) implies that the component $\mathbf{E}_{\parallel}$ of the electric field parallel to $\mathbf{B}$ vanishes. Taking the cross product of Eq. (3.56) with $\mathbf{B}$ and omitting the $\perp$ sign, we get

$$\mathbf{E} \times \mathbf{B} = -(\mathbf{U} \times \mathbf{B}) \times \mathbf{B}$$
$$= -[(\mathbf{U} \cdot \mathbf{B})\mathbf{B} - (\mathbf{B} \cdot \mathbf{B})\mathbf{U}]$$
$$= B^2 \mathbf{U}. \tag{3.58}$$

Since $\mathbf{U} \cdot \mathbf{B} = 0$, one obtains

$$\mathbf{U} = \frac{\mathbf{E} \times \mathbf{B}}{B^2}. \tag{3.59}$$

The above relation implies that in an ideal conducting fluid with large σ, a fluid in motion is equivalent to the presence of an electric field in the rest frame, and vice versa.

Some interesting properties arise when the MHD flow is in a steady state. In this case, $\nabla \times \mathbf{E} = 0$, and so $\mathbf{E}$ can be represented by the electromagnetic potential Φ. For perfectly conducting fluids,

$$\nabla \Phi = \mathbf{U} \times \mathbf{B}. \tag{3.60}$$

If this equation is scalar-multiplied with either $\mathbf{U}$ or $\mathbf{B}$, one gets

$$\mathbf{U} \cdot \nabla \Phi = 0$$
$$\mathbf{B} \cdot \nabla \Phi = 0. \tag{3.61}$$

The above expressions show that $\nabla \Phi$ is perpendicular to both $\mathbf{U}$ and $\mathbf{B}$ and that Φ will be constant along $\mathbf{U}$ and $\mathbf{B}$. The $\Phi = $ constant surfaces are equipotential surfaces and since they are equipotential, the streamlines and the magnetic field lines are also equipotential lines.

Below we give one of the models of the internal structure of a solar prominence as a demonstration of an analytical solution of the MHD equations for the steady-state static configuration. This model in the literature is the famous Kippenhahn–Schluter prominence model. In this model, the temperature is assumed to be a constant $T = T_0$. The width of the prominence is assumed to be much shorter than the height and length, so that one can neglect the variations in the vertical direction and consider only variations in the horizontal direction across the prominence (Nakariakov 2002). Assume

$$\mathbf{B} = (B_{x0}, B_{y0}, B_z(x)), \quad p = p(x), \quad \rho = \rho(x), \tag{3.62}$$

where B_{x0} and B_{y0} are constants. The magnetostatic (being steady and static) equations can be written as

$$0 = -\nabla p = \mathbf{j} \times \mathbf{B} + \rho \mathbf{g}. \tag{3.63}$$

Coupled with

$$\nabla \cdot \mathbf{B} = 0 \tag{3.64}$$

$$\mu \mathbf{j} = \nabla \times \mathbf{B} \tag{3.65}$$

$$p = \frac{\rho RT}{\mu}, \tag{3.66}$$

the horizontal and vertical components of the force balance equation(3.63) are

$$\frac{dp}{dx} = -\frac{B_z}{\mu}\frac{dB_z}{dx} \tag{3.67}$$

$$\frac{B_{x0}}{\mu}\frac{dB_z}{dx} = \rho g. \tag{3.68}$$

Since we have assumed that the temperature is constant, we can use the gas law (3.66) to eliminate the density instead of the pressure and obtain from Eq. (3.68)

$$\frac{B_{x0}}{\mu}\frac{dB_z}{dx} = \frac{p}{\Lambda}, \tag{3.69}$$

where $\Lambda = RT_0/\mu g$ is the pressure scale height. Solving Eq. (3.67), we get

$$p + \frac{B_z^2}{2\mu} = \text{constant.} \tag{3.70}$$

In order to derive the constant, we resort to the boundary conditions, which state that the pressure and the density tend to zero as we move away from the prominence. Also, B_z tends to a constant value B_{z0}, say. Thus,

$$p \to 0 \quad as \quad |x| \to \infty \tag{3.71}$$

$$B_z \to B_{z0} \quad as \quad |x| \to \infty. \tag{3.72}$$

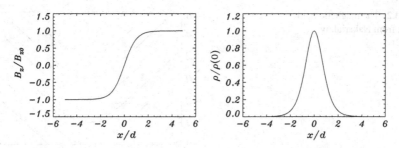

Fig. 3.1 The profiles of B_z and ρ are a function of x/d; from Nakariakov (2002)

Thus,

$$p = \frac{1}{2\mu}(B_{z0}^2 - B_z^2).$$ (3.73)

Substituting the above expression into Eq. (3.69), one gets

$$\frac{B_{x0}}{\mu}\frac{dB_z}{dx} = \frac{1}{2\mu\Lambda}(B_{z0}^2 - B_z^2).$$

Separating the variables and rearranging give

$$\int \frac{dB_z}{(B_{z0}^2 - B_z^2)} = \frac{x}{2\Lambda} + C,$$

where C is a constant. Integrating the left-hand side of the above equation yields

$$\frac{1}{B_{z0}}\tanh^{-1}\left(\frac{B_z}{B_{z0}}\right) = \frac{x}{2\Lambda} + C.$$

which implies that

$$B_z = B_{z0}\tanh\left(\frac{B_{z0}}{2B_{x0}}\frac{x}{\Lambda} + C\right).$$

From the symmetry at $x = 0$, one obtains $B_z(0) = 0$, and this implies that $C = 0$. Thus,

$$B_z = B_{z0}\tanh\left(\frac{B_{z0}}{2B_{x0}}\frac{x}{\Lambda}\right),$$ (3.74)

and the pressure is (see Fig. 3.1)

$$p = \frac{B_{z0}^2}{2\mu}\text{sech}^2\left(\frac{B_{z0}}{2B_{x0}}\frac{x}{\Lambda}\right).$$ (3.75)

Fig. 3.2 The magnetic field lines; from Nakariakov (2002)

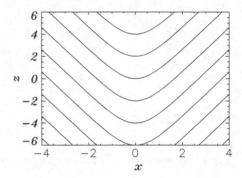

We have assumed that the temperature is constant. From the gas law, one can determine the density as

$$\rho = \frac{\mu}{RT_0} \frac{B_{z0}^2}{2\mu} \operatorname{sech}^2 \left(\frac{B_{z0}}{2B_{x0}} \frac{x}{\Lambda} \right). \tag{3.76}$$

The equation of the field lines for the Kippenhahn–Schluter prominence model is given by

$$\frac{dx}{B_x} = \frac{dz}{B_z}, \tag{3.77}$$

which implies that

$$\int \frac{B_{z0}}{B_{x0}} \tanh \left(\frac{B_{z0}}{2B_{x0}} \frac{x}{\Lambda} \right) dx = z + c.$$

Integrating the above equation, we get

$$2\Lambda \log \left[\cosh \left(\frac{B_{z0}}{2B_{x0}} \frac{x}{\Lambda} \right) \right] = z + c. \tag{3.78}$$

The magnetic field lines given by Eq. (3.78) are plotted in Fig. 3.2. It is interesting to note that the magnetic field lines are bent and that the magnetic tension force opposes the force due to gravity. Also, the magnetic pressure is higher away from the center of the prominence, resulting in the magnetic pressure acting toward the center, which compresses the plasma while opposing the outward pressure gradient.

3.6   MHD Equilibrium

The study of the equilibrium (both static and dynamic) of physical systems is essential to have a better understanding of the evolution of the systems. In particular, in order to study waves and oscillations of a physical system, it is necessary to have an idea about the equilibrium solution of the system. Once the equilibrium state is known, one can perturb the existing equilibrium and study the evolution of such perturbations. These perturbations may be propagating or getting damped as a function of time. Some of the perturbations may involve finite-amplitude waves (which will involve nonlinear wave theory). Others may be stable or unstable also. On related scales of length and time, the assumption of incompressibility is quite valid and useful. There have been several attempts in the past to study the equilibrium of self-gravitating fluids in the presence of magnetic fields, (Chandrasekhar and Fermi 1953; Ferraro 1954; Penderghast 1956; Roberts 1955) which are considered to be very classical but still very useful to study the dynamics of stellar systems, in particular the Sun. Penderghast (1956) showed that in the absence of fluid motions, a spherical equilibrium configuration exists in which the magnetic forces do not vanish. It is important that in any epoch, the steady and fluctuating parts of the rotation and magnetic field must be considered simultaneously. The contribution to the fluctuations could be by superposition of a large number of global solar oscillations (hydromagnetic oscillations) of different dynamical times scales. It may happen that the steady parts of the rotation and magnetic field may themselves vary on very long time scales.

The equilibrium of a self-gravitating incompressible fluid with a magnetic field and large electrical conductivity has been studied by Satya Narayanan (1996). Both the magnetic field and the fluid motion are assumed to have a symmetry about an axis. The basic magnetic field is assumed to be made up of two parts, namely, poloidal and toroidal, while the fluid motion is assumed to be purely rotational. The meridional circulation has been ignored in this case.

In this section, we present two equilibrium solutions (exact) for the magnetohydrostatic and magnetohydrodynamic cases.

The magnetohydrostatic equilibrium is assumed to be (Nakariakov 2002)

$$0 = -\nabla p + \mathbf{j} \times \mathbf{B} + \rho \mathbf{g}, \tag{3.79}$$

along with

$$\nabla \cdot \mathbf{B} = 0 \tag{3.80}$$

$$\mu \mathbf{j} = \nabla \times \mathbf{B} \tag{3.81}$$

$$p = \frac{\rho RT}{\mu}, \tag{3.82}$$

where T, the temperature, satisfies the energy equation. Consider the simple case of a uniform vertical magnetic field that does not exert any force. Also, let's assume that the temperature is known. Then

$$\mathbf{B} = B_0 \hat{z}, \quad \mathbf{g} = -g\hat{z}.$$

For a uniform vertical magnetic field, $\mathbf{j} = 0$ and hence there is no Lorentz force. The relationship between the pressure and the density reduces to

$$\frac{dp}{dz} = -\rho(z)g = -\frac{g\mu}{RT(z)}p(z) = -\frac{p(z)}{\Lambda(z)}, \tag{3.83}$$

where

$$\Lambda(z) = \frac{RT(z)}{\mu g}$$

is the scale height of the pressure. The first-order ordinary differential equation given by Eq. (3.83) is separable and is given by

$$\frac{dp}{p} = -\frac{1}{\Lambda(z)}dz.$$

A straightforward integration of the above equation gives

$$\log p = -n(z) + \log p(0),$$

where

$$n(z) = \int_0^z \frac{1}{\Lambda(u)}du$$

is the integrated number of scale heights between the arbitrary level at which the pressure is $p(0)$ and the height z. Thus,

$$p(z) = p(0)\exp[-n(z)]. \tag{3.84}$$

For an isothermal atmosphere, both T and Λ are constant, so that the above relation reduces to

$$p(z) = p(0)\exp(-z/\Lambda) \qquad \rho(z) = \rho(0)\exp[-z/\Lambda].$$

The pressure has an exponential decay on a typical length scale given by the pressure scale height Λ.

Let's now turn our attention to the axisymmetric MHD equilibrium of a self-gravitating incompressible fluid. The basic equations of MHD for a spherical system can be reduced to that of a set of equations in a cylindrical coordinate system (y, ϕ, z) with a suitable transformation Chandrasekhar (1956). The notation "y" is used instead of the conventional "r" in a cylindrical coordinate system. The hydromagnetic equations for the equilibrium in an incompressible medium with infinite electrical conductivity in which axial symmetry prevails can be written in the form of the system of coupled partial differential equations below. For the scalars P, T, U, V, the first two define the poloidal and toroidal magnetic fields, and the second two the meridional and differential velocity fields.

$$[y^2U, y^2P] = 0 \tag{3.85}$$

$$[y^2U, T] + [V, y^2P] = 0 \tag{3.86}$$

$$[y^2T, y^2O] + [y^2U, y^2V] = 0 \tag{3.87}$$

$$[\triangle_5 P, y^2P] - [\triangle_5 U, y^2U] + y\frac{\partial}{\partial z}(T^2 - V^2) = 0 \tag{3.88}$$

The notation $[F,G]$ stands for the Jacobian of F and G with respect to y and ω; that is,

$$[F,G] = \frac{\partial(F,G)}{\partial(z,\omega)} = \frac{\partial F}{\partial z}\frac{\partial G}{\partial \omega} - \frac{\partial F}{\partial \omega}\frac{\partial G}{\partial z}.$$

Also, $\triangle_5$ is defined as

$$\triangle_5 = \frac{\partial^2}{\partial \omega^2} + \frac{3}{\omega}\frac{\partial}{\partial \omega} + \frac{\partial^2}{\partial z^2}.$$

ω is given by

$$\omega = \frac{p}{\rho} + (1/2)|v|^2 + \mathbf{G},$$

where $\mathbf{G}$ is the gravitational potential.

In stars like the Sun, the meridional motion U is negligible compared to V, so that the equations reduce to (Satya Narayanan 1996)

$$[V,y^2P] = 0 \tag{3.89}$$

$$[y^2T,Y^2P] = 0 \tag{3.90}$$

$$[\triangle_5P,y^2P] + y\frac{\partial T^2}{\partial z} - y\frac{\partial V^2}{\partial z} = 0. \tag{3.91}$$

Equations (3.89) and (3.90) yield

$$V = \text{function} \quad \text{of} \quad (y^2P) \tag{3.92}$$

$$y^2T = \text{function} \quad \text{of} \quad (Y^2P). \tag{3.93}$$

Defining two functions $G(y^2P)$ and $g(Y^2P)$ as

$$G(y^2P) = \frac{1}{2}\frac{d}{d(y^2P)}y^4T^2 \tag{3.94}$$

$$g(y^2P) = \frac{1}{2}\frac{d}{d(y^2P)}V^2. \tag{3.95}$$

Equation (3.91) can be simplified to yield

$$\triangle_5P + \frac{1}{y^2}G(y^2P) + y^2g(y^2P) = \Phi(y^2P). \tag{3.96}$$

The above equation represents a general integral of the equilibrium solution for the case $U = 0$. The above equation is highly nonlinear in P and in general does not admit closed-form solutions. Hence, one may have to resort to numerical procedures by specifying forms of G, g, and Φ. It is interesting to note that it is possible to reduce this equation to a linear one by combining the choice of G, g, and Φ. The simplest linear equation is thus obtained by choosing

$$G(y^2 P) = \alpha^2 y^2 P$$

$$g(y^2 P) = \beta/2$$

$$\Phi(y^2 P) = k, \tag{3.97}$$

where α, β and k are constants. With the above choice, Eq. (3.96) reduces to

$$\triangle_5 P + \alpha^2 P = k - \beta y^2/2. \tag{3.98}$$

Let's try to find an exact solution of the above equation. For α different from 0, the solution of the homogeneous part of the above equation is given by

$$P = \sum_{n=0}^{\infty} A_n \frac{J_{n+3/2}(\alpha r)}{(\alpha r)^{3/2}} C_n^{3/2}(\mu). \tag{3.99}$$

$J_{n+3/2}(\alpha r)$ is the Bessel function of order $(n+3/2)$ and $C_n^{3/2}(\mu)$ is the Gegenbauer polynomial of order n and index $3/2$.

The particular integral can be found by assuming a power series in y and z in the form

$$P_1 = \sum_{n,m=0}^{\infty} a_{nm} y^n z^m. \tag{3.100}$$

Substituting the above expression into the equation, by simple algebra one can show that

$$P_1 = \frac{k}{\alpha^2} + \frac{4\beta}{\alpha^4} - \frac{\beta}{2\alpha^2} y^2.$$

Thus, the general solution in spherical polars can be written as

$$P = \sum_{n=0}^{\infty} A_n \frac{J_{n+3/2}(\alpha r)}{(\alpha r)^{3/2}} C_n^{3/2}(\mu) + \frac{k}{\alpha^2} + \frac{4\beta}{\alpha^4} - \frac{\beta}{2\alpha^2} r^2 (1 - \mu^2). \tag{3.101}$$

The solution for the poloidal magnetic field when it is current-free is given by

$$P = \sum_{n=0}^{\infty} \frac{A}{r^{n+3}} C_n^{3/2}(\mu). \tag{3.102}$$

With the choice given by Eq. (3.97), the expressions for V and T become

$$V^2 = \beta y^2 P + \gamma$$

and

$$T^2 = \alpha^2 P^2 + \frac{\delta}{y^4}, \tag{3.103}$$

where α, β, γ, and δ are all constants. It is easily seen that T is regular on the axis $(y = 0)$ only if $\delta = 0$. Thus, we have

$$T = \alpha P$$

and

$$V = (\beta y^2 P + \gamma)^{1/2}. \tag{3.104}$$

Relation (3.101) with (3.104) gives a simple, exact form of the equilibrium solution of the MHD equations. The above relations are only a restricted class of solutions of the general integral (3.96) of the equilibrium solutions.

3.7 Magnetic Flux Tubes

Magnetic configurations are made up of two types, the magnetic flux tube and a current sheet. In this section, the basic properties of flux tubes (without proving them) will be presented. One of the most important examples of a flux tube in the Sun is the sunspot, observed in the photosphere, where a large magnetic tube breaks through the solar surface. Erupting prominences may be considered monolithic flux tubes. Recent observations of coronal loops reveal that they too have flux tube structures.

A magnetic field line may be defined as the tangent at a given point in the direction of the field **B** (Priest 1982). In the Cartesian coordinates (for two dimensions), the magnetic lines are the solution of the following equation:

$$\frac{dy}{dx} = \frac{B_y}{B_x}.$$

In three dimensions, the equations become

$$\frac{dx}{B_x} = \frac{dy}{B_y} = \frac{dz}{B_z}.$$

B_x and so on are the magnetic field components in the x, y, z-directions, respectively. The definition for the magnetic field (or flux) tube is that it is the volume enclosed by the set of field lines that intersect a simple closed curve in space. This leads to the definition of the strength of the flux tube, which states that it is the amount of flux crossing a section S. In terms of an integral, it is simply

$$F = \int_S \mathbf{B} \cdot d\mathbf{S}, \tag{3.105}$$

where $d\mathbf{S}$ and **B** have the same sense so that F takes positive values.

General properties: (1) The strength of a flux tube remains constant along its length. (2) The strength of a flux tube increases when it narrows, while it decreases

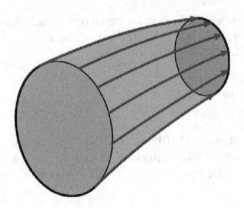

when it widens. (3) The magnetic field B and density ρ increase in the same proportion when the tube is compressed. (4) An extension of a flux tube increases the field strength when it is not compressed. (5) A twist in the flux tube when not tethered is unstable and is referred to as the helical kink instability. (6) The behavior of the normal modes in a uniform plasma is modified by the geometry when they propagate along a flux tube.

We will have occasion to deal more with the nature of the modes that will appear in a flux tube in Chap. 5, where we will deal with oscillations in cylindrical geometry. A cylinder is a special case of a flux tube.

The concept of flux freezing is an important property that occurs when the magnetic Reynolds number $R_m \gg 1$. In this case, the diffusion term in the magnetic induction equation can be neglected and one considers the ideal MHD with the following equation:

$$\frac{\partial \mathbf{B}}{\partial t} = \nabla \times (\mathbf{U} \times \mathbf{B}). \tag{3.106}$$

One can show from that above equation that (reference will be provided)

$$\frac{\mathrm{d}}{\mathrm{d}t} \int_S \mathbf{B} \cdot \mathrm{d}\mathbf{S} = 0. \tag{3.107}$$

The physical interpretation of the above equation is that if a magnetic field vector in a plasma system satisfies Eq. (3.106), then the magnetic flux through any surface (say S) constituting a part of the moving fluid will remain time-invariant when that fluid element is moving. That is, the magnetic field remains frozen-in to the flow and moves along with it. This is the famous flux-freezing theorem due to Alfvén. An important consequence of the above result is that any twisting or stretching motion in a magnetized plasma will result in the magnetic field being twisted or stretched.

3.8 Current-Free (Potential) Fields

An interesting situation is when the current density vanishes everywhere, with the neglect of gas pressure (with respect to the magnetic pressure), a situation where the plasma β is very small. The magnetostatic equation in this case reduces to

$$\mathbf{j} \times \mathbf{B} = 0. \tag{3.108}$$

With the assumption that the current density is identically zero, the magnetic field is potential. In this case, the field must satisfy the conditions

$$\mathbf{j} = \nabla \times \mathbf{B} = 0 \tag{3.109}$$

and

$$\nabla \cdot \mathbf{B} = 0. \tag{3.110}$$

The most general solution to Eq. (3.109) is

$$\mathbf{B} = \nabla \phi, \tag{3.111}$$

where ϕ is the scalar magnetic potential. Substituting the above expression into the magnetic divergence-free condition, we get

$$\nabla^2 \phi = \frac{\partial^2 \phi}{\partial x^2} + \frac{\partial^2 \phi}{\partial y^2} + \frac{\partial^2 \phi}{\partial z^2}. \tag{3.112}$$

The above equation is the famous Laplace equation, whose solution can be found in any textbook on mathematical physics or special functions. However, for the sake of completeness, we shall present the solution.

We shall solve the equation in two dimensions, the (x,y)-plane, subject to the boundary conditions

$$\phi(x,0) = F(x), \quad \phi(0,y) = \phi(l,y) = 0, \quad \phi \to \quad as \quad y \to \infty. \tag{3.113}$$

Substituting $\phi = X(x)Y(y)$ into Eq. (3.112) gives

$$X''Y + XY'' = 0, \tag{3.114}$$

which can be simplified to yield

$$\frac{X''}{X} = -\frac{Y''}{Y} = \text{constant} = -k^2. \tag{3.115}$$

The above expression is equivalent to two ordinary differential equations:

$$Y'' = k^2 Y \implies Y(y) = ae^{-ky} + be^{ky}$$

and

$$X'' = -k^2 X, \implies X(x) = c\sin(kx) + d\cos(kx). \tag{3.116}$$

Applying the boundary conditions, one obtains $b = d = 0$ and

$$\sin(kl) = 0 \quad \Longrightarrow k = \frac{n\pi}{l}.$$

Summing all the possible solutions, the solution for ϕ may be obtained. Defining $A_k = ac$, we get

$$\phi(x,y) = \sum_k A_k \sin(kx) e^{-ky}, \tag{3.117}$$

where $F(x) = \sum_k A_k \sin(kx)$ and $k = n\pi/l$. For the sake of brevity, let's assume that $F(x)$ has only one Fourier component, namely, $F(x) = \sin(\pi x)/l$. The potential solution is simply given by

$$\phi(x,y) = \sin(\pi x/l) e^{-\pi y/l}. \tag{3.118}$$

Once we know ϕ, we can calculate the components of the magnetic field by using the relation $\mathbf{B} = \nabla\phi$:

$$B_x = \frac{\partial \phi}{\partial x} = B_0 \cos\left(\frac{\pi x}{l}\right) e^{-\pi y/l} \tag{3.119}$$

$$B_y = \frac{\partial \phi}{\partial y} = -B_0 \sin\left(\frac{\pi x}{l}\right) e^{-\pi y/l}. \tag{3.120}$$

The solution of the Laplacian mentioned above is in terms of the Cartesian coordinates. However, one can write the general solution in terms of spherical polar coordinates (r, θ, ϕ), and it is given by

$$\phi = \sum_{l=0}^{\infty} \sum_{m=-l}^{l} (a_{lm} r^l + b_{lm} r^{-(l+1)}) P_l^m(\cos\theta) e^{im\phi}, \tag{3.121}$$

where P_l^m is the associated Legendre polynomial. In cylindrical polar coordinates (R, ϕ, z), the general solution may be written in terms of

$$\phi = \sum_{n=-\infty}^{\infty} (c_n J_n(kR) + d_n Y_n(kR)) e^{in\phi \pm kz}, \tag{3.122}$$

where J_n and Y_n are the Bessel functions. In each of the cases mentioned above, the arbitrary constants can be calculated by applying suitable boundary conditions.

Application: A coronal current-free field model, where the asymptotic condition of no field at infinity is applicable and the boundary condition on the solar surface has been specified, has been studied by Yan (2005). The method applied to practical solar events indicates that the extrapolated global magnetic field structures effectively demonstrate the case for the disk signature of the radio CME and the evolution of the radio sources during the CME/flare processes.

The magnetic field has been represented as a scalar potential ψ with $\mathbf{B} = -\nabla\psi$. The Laplace equation can be written as

$$\nabla^2 \psi = 0.$$

On the surface of the Sun, say S, the line-of-sight field component or the normal component B_n is specified as

$$B_n = -\frac{\partial \psi}{\partial n}.$$

In general, the potential ψ at any position $\mathbf{r_i}$ in a volume V can be determined as

$$\psi(\mathbf{r_i};\mathbf{r}) = \int_S \left[G(\mathbf{r_i};\mathbf{r}) \frac{\partial \psi(\mathbf{r})}{\partial \mathbf{n}} - \frac{\partial G(\mathbf{r_i};\mathbf{r})}{\partial \mathbf{n}} \psi(\mathbf{r}) \right] d\mathbf{S},$$

where $G = 1/4\pi|\mathbf{r_i} - \mathbf{r}|$ is Green's function of the Laplace equation in free space.

The magnetic field can be derived from the formula

$$\mathbf{B} = \int_S \left[\psi \frac{\partial}{\partial r} \left(\frac{\partial G}{\partial n} \right) - \frac{\partial \psi}{\partial n} \frac{\partial G}{\partial r} \right] ds.$$

3.9 Force-Free Fields

For a plasma β whose value is much smaller than unity, the Lorentz force dominates the pressure gradient and the gravitational forces. With this assumption, the magnetohydrostatic equation simplifies to

$$\mathbf{j} \times \mathbf{B} = 0. \tag{3.123}$$

This implies that the electric current flows along the magnetic field lines. Combining Ampere's law with the above expression, we obtain

$$\nabla \times \mathbf{B} = \alpha \mathbf{B}, \tag{3.124}$$

where α is in general some function of the position. The only restriction for α is that it is a constant along each magnetic field line. Taking divergence of the above equation results in

$$(\mathbf{B} \cdot \nabla)\alpha = 0, \tag{3.125}$$

which implies that $\mathbf{B}$ lies on surfaces of constant α. The field is said to be a linear or constant α field, when α takes the same value on each field line. Taking the curl of Eq. (3.124) results in a linear equation:

$$(\nabla^2 + \alpha^2)\mathbf{B} = 0. \tag{3.126}$$

This is the famous Helmholtz equation. For nonconstant α, the governing equation is nonlinear and we will mention it in subsequent pages. More attention has been paid to the constant α-field, due to the difficulty in finding general solutions to Eq. (3.123), although it looks very simple.

We shall mention the famous Woltjer theorem Priest (1982a) without proof that "for a perfectly conducting plasma in a closed volume (V_0), the integral

$$\int_{V_0} \mathbf{A} \cdot \mathbf{B} dV = K_0 \tag{3.127}$$

is invariant and the state of minimum magnetic energy is a linear (constant α) magnetic field and force-free magnetic field"). Here $\mathbf{A}$ is a vector potential, as defined in Chap. 2, given by $\mathbf{B} = \nabla \times \mathbf{A}$. The proof of the above theorem can be obtained through a variational approach.

Simple constant α-solutions: The form of the simple solution is

$$\mathbf{B} = (0.B_y(x), B_z(x)),$$

which is one-dimensional. The differential equation reduces to

$$\frac{d}{dx}\left(B_y^2 + B_z^2\right) = 0.$$

The divergence of the magnetic field is satisfied identically. Integrating the above equation with respect to x gives

$$B_y^2 + B_z^2 = B_0^2,$$

and the solution can be written as

$$\mathbf{B} = \left(0, B_y, \left(B_0^2 - B_y^2\right)^{1/2}\right).$$

For the constant α-case, the z-component of Eq. (3.124) is given by

$$\frac{dB_y}{dx} = \alpha \left(B_0^2 - B_y^2\right)^{1/2}.$$

With the choice of the origin to be a root of B_y, the solution can be written as

$$B_y = B_0 \sin(\alpha x), \quad B_z = B_0 \cos(\alpha x). \tag{3.128}$$

One of the simple two-dimensional solutions in separable form is given by

$$B_x = A_1 \cos(kx) e^{-lz}$$
$$B_y = A_2 \cos(kx) e^{-lz}$$
$$B_z = B_0 \sin(kx) e^{-lz}. \tag{3.129}$$

The corresponding solution in cylindrical coordinates (R, ϕ, z) is given by

$$B_R = (l/k) B_0 J_1(kR) e^{-lz}$$
$$B_\phi = (1 - l^2/k^2)^{1/2} B_0 J_1(kR) e^{-lz}$$
$$B_z = B_0 J_0(kR) e^{-lz}. \tag{3.130}$$

The general solution in spherical polar coordinates (r, θ, ϕ) can be written as

$$B_r = C_n n(n+1) r^{-3/2} J_{n+1/2}(\alpha r) P_n(\cos\theta)$$

$$B_\theta = C_n[-nr^{-3/2} J_{n+1/2}(\alpha r) + \alpha r^{-1/2} J_{n-1/2}(\alpha r)] \frac{dP_n(\cos\theta)}{d\theta}$$

$$B_\phi - C_n \alpha r^{-1/2} J_{n+1/2}(\alpha r) \frac{dP_n(\cos\theta)}{d\theta}. \tag{3.131}$$

General constant α-solutions: The basic differential equation (Priest 1982)

$$(\nabla^2 + \alpha^2)\mathbf{B} = 0 \tag{3.132}$$

requires that

$$\nabla \cdot \mathbf{B} = 0.$$

The general form of $\mathbf{B}$ may be assumed to be

$$\mathbf{B} = \alpha \nabla \times (\psi\mathbf{a}) + \nabla \times (\nabla \times (\psi\mathbf{a})), \tag{3.133}$$

where $\mathbf{a}$ is a constant vector and the scalar function ψ satisfies the Helmholtz equation,

$$(\nabla^2 + \alpha^2)\psi = 0. \tag{3.134}$$

For the specific case of $\mathbf{a} = \hat{\mathbf{z}}$, the solution of the above equation in Cartesian coordinates with the condition that each term tends to zero as $z \to \infty$ is given by

$$\psi = \int_0^\infty \int_0^\infty A(k_x, k_y) e^{i\mathbf{k}\cdot\mathbf{r} - lz} dk_x dk_y,$$

where $\mathbf{k} = k_x\hat{\mathbf{x}} + k_y\hat{\mathbf{y}}$, $l = (k^2 - \alpha^2)^{1/2}$, and $A(k_x, k_y)$ are the complex constants ($k \neq 0$). The magnetic field components can be written as

$$B_x = \int_0^\infty \int_0^\infty i(\alpha k_y - lk_x) A(k_x, k_y) e^{i\mathbf{k}\cdot\mathbf{r} - lz} dk_x dk_y$$

$$B_y = -\int_0^\infty \int_0^\infty i(\alpha k_y + lk_x) A(k_x, k_y) e^{i\mathbf{k}\cdot\mathbf{r} - lz} dk_x dk_y$$

$$B_z = \int_0^\infty \int_0^\infty k^2 A(k_x, k_y) e^{i\mathbf{k}\cdot\mathbf{r} - lz} dk_x dk_y. \tag{3.135}$$

For the case $\mathbf{a} = \mathbf{r}$, the general solution of the force-free equation in spherical polars is given by the famous Chandrasekhar and Kendall (1957) function as follows:

$$\psi = \sum_{n=0}^\infty \sum_{m=0}^\infty A_n^m r^{-1/2} J_{n+1/2}(\alpha r) P_n^m(\cos\theta) e^{im\phi}$$

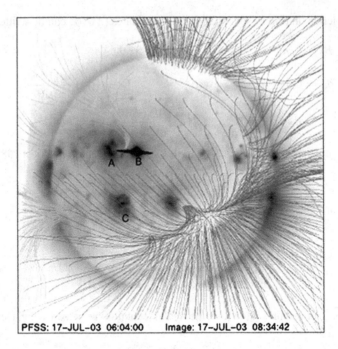

PFSS: 17–JUL–03 06:04:00   Image: 17–JUL–03 08:34:42

Fig. 3.4 Open field lines calculated with the PFSS model; from Nitta and DeRosa (2008)

in terms of the Bessel functions (J_n) and the associated Legendre functions $(P_n^m(\cos\theta))$. Let's now turn to situations in hydrodynamics where one comes across the Helmholtz equation. The first example would be for two-dimensional incompressible flows with rotation:

Application: In a recent paper, Liu et al. (2011) compared a potential field extrapolation to three nonlinear force-free (NLFF) field extrapolations [optimization, direct boundary integral (DBIE), and approximate vertical integration (AVI) methods] to study the spatial configuration of the magnetic field in the quiet Sun. They found that the differences in the computed field strengths among the three NLFF and potential fields exist in the low layers. However, they tend to disappear as the height increases, and the differences are of the order of 0.1 gauss when the height exceeds $\approx 2,000$ km above the photosphere. The difference in azimuth angles between each NLFF field model and the potential field is as follows: For the optimization field, it decreases evidently as the height increases; for the DBIE field, it almost stays constant and shows no significant change as the height increases; for the AVI field, it increases slowly as the height increases. Their analysis show that the reconstructed NLFF fields deviate significantly from a potential field extrapolation and three nonlinear force-free (NLFF) fields (Fig. 3.4).

The potential field source surface (PFSS) model is often employed to extrapolate the photospheric magnetic field to the corona for heliophysics research and applications. Nitta and DeRosa (2008) attempted to evaluate the performance of the

PFSS model, by comparing the computed footpoints of the heliospheric magnetic field with the locations of flares associated with type III radio bursts, which are a good indicator of open field lines that extend to interplanetary space. They discuss possible reasons for the discrepancy, including the model's inadequacy to reproduce the coronal magnetic field above evolving active regions and the lack of a simultaneous full-surface magnetic map. They concluded that the performance of the PFSS model needs to be quantified further against solar observations, including type III bursts, before applying it to heliospheric models.

For an incompressible fluid,

$$\nabla \cdot \mathbf{V} = 0, \tag{3.136}$$

where $\mathbf{V}$ is the velocity field. Let's assume the density ρ to be uniform so that the momentum equation for a uniformly rotating fluid with gravity can be written as

$$\frac{\partial \mathbf{V}}{\partial t} + (2\Omega + \omega) = -\frac{\nabla p}{\rho} + \nabla \left[\Phi - \frac{\mathbf{V}^2}{2} \right] + \frac{F}{\rho}. \tag{3.137}$$

It can be shown by simple calculations that for a uniformly rotating fluid (like a rigid body), the vorticity ω and the angular velocity Ω have the following relationship:

$$\omega = 2\Omega.$$

By simple algebraic simplification, it can be shown that the potential vorticity $(\omega_a = \omega + 2\Omega)$ is a conserved quantity Hasegawa 1985). The term due to gravity is in the form of a gravitational potential, which can be included with the pressure gradient term. Hasegawa (1985) described the general features of a system capable of exhibiting self-organization. The system is described by a nonlinear partial differential equation with dissipation. The system has two or more quadratic or higher-order conserved quantities in the absence of dissipation. If the spectral behavior of these invariants in the inertial range is such that one of them transfers toward large spatial scales and the other to small spatial scales, they would have differential dissipation rates when the dissipation is introduced. The importance of mean square helicity as an invariant and its role in inverse cascade in three-dimensional hydrodynamic flows were pointed out by Levich and Tzevetkov (1985). A variational principle connecting the invariants E (energy) and I (mean square helicity density) for a three-dimensional incompressible fluid, similar to the arguments of Hasegawa (1985), was studied by Satya Narayanan (1993). A closed-form solution of the simplified variational equation was presented.

The variational equation is set up as

$$\delta \int (2\Omega + \nabla \times V)^2 dV - \lambda \delta \int V^2 dV = 0. \tag{3.138}$$

If the boundary is such that the vorticity ω vanishes at the boundary (or periodic boundary condition), the above relation can be simplified with the introduction of a stream function $\psi(x,y)$ as

$$\nabla^2 \psi + \lambda \psi = 0. \tag{3.139}$$

This is the famous Helmholtz equation, whose solution can be written as

$$\psi = \psi_0 \cos\left[\frac{2\pi x}{a}\right] \sin\left[\frac{2\pi y}{b}\right], \qquad (3.140)$$

where a and b are the length and breadth of the rectangular box under consideration, respectively. The above self-organized state will be a stationary solution of the dynamical equation.

For the three-dimensional hydrodynamical flows without rotation, the helicity density γ, a measure of the knottedness of the vorticity field, is defined as $\gamma = \mathbf{V} \cdot \omega$, where $\mathbf{V}$ and ω are the velocity and vorticity, respectively. The quantity I defined as

$$I = \int \, <\gamma(x)\gamma(x+r)> \, d^3r,$$

where $<>$ denote an average over an ensemble, which is an invariant for an ideal 3D hydrodynamical system in addition to the total energy E. Setting up a variational formulation, the equation for the three-dimensional system can be written as

$$2\omega(\mathbf{V} \cdot \omega) - \lambda \mathbf{V} - \mathbf{V} \times \nabla(\mathbf{V} \cdot \omega) = 0. \qquad (3.141)$$

The corresponding equation for a two-dimensional system with enstrophy and energy as its invariants is given by

$$\nabla \times \nabla \times \mathbf{V} - \alpha \mathbf{V} = 0.$$

The above equation can easily be solved by introducing a stream function. The equation for the three dimensions is rather complicated. However, if we consider a quasi-two-dimensional flow, then one can write the solution explicitly. Let's assume the velocity to be

$$\mathbf{V} = \mathbf{V}[V_x(x,y), V_y(x,y), V_z],$$

where V_z is a constant. The variational equation can be recast as

$$\left(\nabla^2 + \frac{\lambda}{V_z^2}\right) \mathbf{V}_H = 0, \qquad (3.142)$$

where $\mathbf{V}_H = (V_x, V_y)$ and $\nabla^2 = \partial^2/\partial x^2 + \partial^2/\partial y^2$. The solution for the above equation can be written as

$$V_x = \hat{V}\cos(2\pi X)\cos(2\pi Y)$$
$$V_y = \hat{V}(b/a)\sin(2\pi X)\sin(2\pi Y).$$

Here $\hat{V}$, a,b are constants. X and Y are given by $X = x/a$ and $Y = y/b$. For the quasi-two-dimensional case, the velocity and vorticity can be written in terms of a stream function ψ as

$$\mathbf{V} = -\nabla \psi \times \hat{z} + V_z \hat{z}$$

$$\omega = \nabla^2 \psi \hat{z},$$

where $\hat{z}$ is the unit vector along the z-direction. In terms of the stream function ψ, the vorticity equation can be written as

$$\frac{\partial}{\partial t}\nabla^2 \psi \hat{z} + [-\nabla \psi \times \hat{z} + V_z \hat{z}] \cdot \nabla(\nabla^2 \psi \hat{z}) - \nu \nabla^4 \psi = 0. \qquad (3.143)$$

An exact solution of the above equation can be written as

$$\psi(X,Y,t) = \hat{\psi}[1 + \exp(-\nu \hat{\lambda} t)]\cos(2\pi X)\sin(2\pi Y).$$

Here $\hat{\psi}$, a, b are constants. $\hat{\lambda} = \lambda/V_z^2$, V_x, V_y and ω can be written as

$$V_x = \hat{V}[1 + \exp(-\nu \hat{\lambda} t)]\cos(2\pi X)\cos(2\pi Y)$$

$$V_y = \hat{V}(b/a)[1 + \exp(-\nu \hat{\lambda} t)]\sin(2\pi X)\sin(2\pi Y)$$

$$\omega = \hat{\omega}[1 + \exp(-\nu \hat{\lambda} t)]\cos(2\pi X)\sin(2\pi Y).$$

In the limit $\nu \to 0$, the above solution represents the self-organized solution of the variational equation for the quasi-two-dimensional hydrodynamic flows. In the limit of the helicity γ, being a constant, the variational equation reduces to (Satya Narayanan et al. 2004)

$$\omega = \alpha \mathbf{V},$$

which is the famous "Beltrami equation" in hydrodynamics whose solutions are given by

$$u = \cos(y) - \sin(z)$$

$$v = \cos(z) - \sin(x)$$

$$w = \cos(x) - \sin(y).$$

Nonconstant α-solutions: For the case when α is not a constant, the equation (magnetohydrostatic) is more difficult to solve. The governing equation would be

$$\mathbf{j} \times \mathbf{B} = 0,$$

where $\mathbf{j} = \nabla \times \mathbf{B}/\mu$ and the divergence of $\mathbf{B}$ is zero. We have

$$\nabla \times (\nabla \times \mathbf{B}) = \nabla \times (\alpha \mathbf{B}) = \alpha \nabla \times \mathbf{B} + \nabla \alpha \times \mathbf{B}$$

$$= \alpha^2 \mathbf{B} + \nabla \alpha \times \mathbf{B}.$$

Thus, the Helmholtz equation is modified as

$$\nabla^2 \mathbf{B} + \alpha^2 \mathbf{B} = \mathbf{B} \times \nabla \alpha,$$

with an additional equation given by

$$\mathbf{B} \cdot \nabla \alpha = 0.$$

These equations are very complicated to solve analytically and are usually solved numerically.

3.10   Parker's Solution for Solar Wind

In 1958, motivated by diverse indirect observations, E. N. Parker (1958) developed the first fluid model of a continuously expanding solar corona driven by the large pressure difference between the solar corona and the interstellar plasma. His model produced low flow speeds close to the Sun, supersonic flow speeds far from the Sun, and vanishingly low pressures at large heliocentric distances. In view of the fluid character of the model, he called this continuous supersonic expansion the solar wind. The electrical conductivity of the solar wind plasma is so high that the solar magnetic field is frozen into the solar wind flow as it expands outward from the Sun. Because the Sun rotates with an average period of 27 days, the magnetic field lines in its equatorial plane are bent into spirals whose inclination to the radial direction depend on heliocentric distance and the speed of the wind. At 1 AU, the average field is inclined $\approx 45°$ to the radial direction in the equatorial plane. Measurements made by instruments on board Mariner II during its epic three-month journey to Venus in 1962 provided firm confirmation of a continuous solar wind flow and spiral heliospheric magnetic field that agree with Parker's model, on average. Mariner II also showed that the solar wind is highly variable, being structured into alternating streams of high- and low-speed flows that last for several days each. The observed magnetic field was also highly variable in both strength and orientation. Solar rotations produce radial variations in speed. Faster wind overtakes slow wind ahead while outrunning slow wind behind. As a result, the leading edges of high-speed streams steepen with increasing heliocentric distance. Plasma is compressed on the leading edge of a stream and rarefied on the trailing edge. The buildup of pressure on the leading edge of a stream produces forces that accelerate the low-speed wind ahead and decelerate the high-speed wind within the stream. When the difference in speed between the crest of a stream and the trough ahead is greater than about twice the sound speed, ordinary pressure signals do not propagate fast enough to move the slow wind out of the path of the fast wind, and a forward-reverse shock pair forms on the opposite sides of the high-pressure region. Although the shocks propagate in opposite directions relative to the solar wind, both are carried away from the Sun by the high-bulk flow of the wind. The major accelerations and decelerations of the wind then occur at the shocks, and the stream profile becomes a damped double sawtooth. Because the sound speed decreases with increasing heliocentric distance, virtually all high-speed streams eventually have shock pairs on their leading edges. The dominant structure in the solar equatorial plane in the outer heliosphere is the expanding compression regions where most of the plasma and magnetic field are concentrated (Fig. 3.5).

Fig. 3.5 Solution of the solar wind equation (3.149). The physically relevant solution starts at a low velocity near the solar surface, passes through the critical point at $\frac{r}{r_c} = 1$ and $\frac{v}{v_c} = 1$, and becomes supersonic at large distances; from Nakariakov (2002)

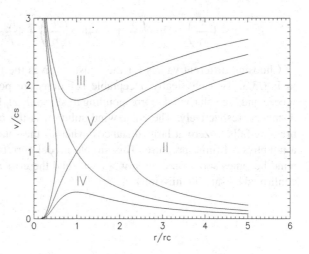

Below we give the theoretical description of Parker's model for the solar wind.

The basic assumptions in the solar wind model are that the corona is very hot, while the interstellar space is very cold. The corona is not in static equilibrium, and it is continually expanding outward, with an outflow of the plasma from the Sun. Assuming a steady, spherically symmetric flow, the MHD equations can be written as follows: Mass continuity reduces to

$$\frac{d}{dr}(r^2 \rho v) = 0; \tag{3.144}$$

the momentum equation is

$$\rho v \frac{dv}{dr} = -\frac{dp}{dr} - \rho g R_\odot^2; \tag{3.145}$$

the ideal gas law is

$$p = \rho R T; \tag{3.146}$$

and the energy equation is

$$T = \text{constant}, \tag{3.147}$$

where v is the radial velocity, ρ the density, p the pressure, g the surface gravitational acceleration, T the temperature, r the radius, and $R_\odot$ the solar radius. Equations (3.144)–(3.147) can be rearranged in terms of the radial velocity as

$$\left(v - \frac{v_c^2}{c}\right)\frac{dv}{dr} = 2\frac{v_c^2}{r} - g\frac{R_\odot^2}{r^2}, \tag{3.148}$$

where $v_c^2 = RT$ is the square of the isothermal sound speed. Equation (3.148) has a singularity at the sonic point $v = v_c$ and $r = r_c$, where the critical radius is $r_c = gR_\odot^2/2v_c^2$. Solving Eq. (3.148) gives the classic Parker solar wind solution, namely,

$$\left(\frac{v}{v_c}\right)^2 - \log\left(\frac{v}{v_c}\right)^2 = 4\log\left(\frac{r}{r_c}\right) + 4\frac{r_c}{r} + \text{constant}. \qquad (3.149)$$

Choosing different values for the constant gives the family of solutions shown in Fig. 3.3. Two physically acceptable solutions are possible, namely, the solar breeze and the solar wind, corresponding to subsonic and supersonic flows at large distances, respectively. The solar wind solution is the correct solution, since the pressure falls to zero at large distances, whereas the solar breeze solution predicts a nonphysical finite pressure. This simple model correctly predicts that the solar wind becomes supersonic, with $v > v_c$ beyond the critical radius r_c, and has been confirmed by satellite missions.

Chapter 4
Waves in Uniform Media

4.1 Basic Equations

Waves and oscillations are ubiquitous in nature when the state of matter is either a liquid or gas. The Sun, being a huge ball of ionized gas, is no exception. The study of waves in the solar atmosphere started with the observation of Doppler shifts (line spectra being wiggly) and the broadening of the spectral lines of the photosphere. It was conjectured that the acoustic waves, generated in the hydrogen-rich convection zone, supplied the nonradiative energy to heat the chromosphere and the corona. The main restoring forces that act on a gas in the atmosphere of the Sun are (1) gas pressure, (2) gravity, and (3) magnetic fields. A gas that is in a stable equilibrium will start oscillating when disturbed. If, for example, only one of the above forces acts, the resulting characteristic oscillations, or modes, are termed *sound*, *internal gravity*, or *Alfvén waves*, respectively. The basic physics of the modes are relatively simpler to understand; however, the properties of the oscillations when two or more forces act simultaneously on a gas have a very complicated dependence on the period and wavelength.

In the solar atmosphere, from the photosphere to the corona, structures due to the magnetic field and stratification due to gravity are present. In such a medium, the propagation of magnetohydrodynamic (MHD) waves is rather complicated, with the well-known results of the uniform medium providing limited information on the behavior of the modes in the inhomogeneous atmosphere of the Sun. However, some basic knowledge of the behavior of these modes in a homogeneous medium will shed some light on how the inhomogeneous medium may behave qualitatively.

The Sun, being a dynamic body, contains features that are continually changing over a wide range of scales (spatial and temporal). One interpretation of the waves, which propagate outward from the umbra in the sunspots, may be due to fast magnetoacoustic gravity waves. Also, the aftermath of a flare may lead to a Moreton (or flare-induced coronal) wave (named after the person who suggested it) in the form of a magnetoacoustic wave. More discussion on Moreton waves will be presented in Chap. 7. Both the photosphere and the chromosphere exhibit small-scale

A. Satya Narayanan, *An Introduction to Waves and Oscillations in the Sun*, Astronomy and Astrophysics Library, DOI 10.1007/978-1-4614-4400-8_4,
© Springer Science+Business Media New York 2013

wave motions, with a period of about 300 s, the famous 5-min oscillations. The different types of waves discussed above are far from a complete list. In what follows in this chapter, we shall discuss the different types of waves possible in a uniform media. By uniform media, we mean a situation where the density, pressure, or magnetic field may vary with respect to the vertical height or radially, depending on the geometry chosen. However, no discontinuities in the density, pressure, or magnetic field are present in the media.

To begin with, we shall derive the dispersion relation for different types of modes using the standard theory of waves by linearizing the basic equations of motion. In this we consider a basic equilibrium situation, perturb it slightly, and assume that the disturbance propagates in the form of a wave. The linearized equations of motion are used along with the perturbed quantities, which vary, such as $\exp[i(\mathbf{k} \cdot \mathbf{r} - \omega t)]$, to derive the dispersion relation, which relates the frequency ω and the wavenumber $\mathbf{k}$.

The basic equations of motion describing the ideal MHD fluid are the continuity of mass, momentum, and energy, together with the induction equation, as follows (Priest 1982):

$$\frac{D\rho}{Dt} + \rho \nabla \cdot \mathbf{v} = 0 \tag{4.1}$$

$$\rho \frac{D\mathbf{v}}{Dt} = -\nabla p + (\nabla \times \mathbf{B}) \times \mathbf{B}/\mu - \rho g \hat{\mathbf{z}} - 2\rho \Omega \times \mathbf{v} \tag{4.2}$$

$$\frac{D}{Dt}\left(\frac{p}{\rho^\gamma}\right) = 0 \tag{4.3}$$

$$\frac{\partial \mathbf{B}}{\partial t} = \nabla \times (\mathbf{v} \times \mathbf{B}) \tag{4.4}$$

$$\nabla \cdot \mathbf{B} = 0. \tag{4.5}$$

The electric current and temperature have the following forms:

$$\mathbf{j} = \nabla \times \mathbf{B}/\mu \tag{4.6}$$

$$T = \frac{mp}{k_B \rho}. \tag{4.7}$$

Here, ρ is the density, $\mathbf{v}$ the velocity, p the pressure, $\mathbf{B}$ the magnetic induction, g the acceleration due to gravity, Ω the angular velocity, μ the magnetic permeability, and γ the ratio of specific heats, respectively.

For simplicity, we shall work in a rotating frame with the Sun, whose angular velocity (Ω) is assumed to be constant relative to an inertial frame. The effect of rotation does not produce a significant effect on Maxwell's equation if the absolute speed $(\Omega \times \mathbf{r} + \mathbf{v})$ is nonrelativistic. Rotation, as we all know, gives rise to the Coriolis force, given by $(-2\rho\Omega \times \mathbf{v})$, together with a centrifugal force, $(1/2\rho\nabla[\Omega \times \mathbf{r}]^2)$, not written explicitly in the equations of motion as it can be combined with the gravitational term. The gravitational force $-\rho g \hat{\mathbf{z}}$ is assumed to be constant, with the z-axis directed along the outward normal to the surface of the Sun. The ratio $p : \rho^\gamma$ is assumed to be constant following the motions as the plasma is frozen to the magnetic field and thermally isolated from the surroundings.

Let's discuss the equilibrium situation before linearizing the equations and assuming wavelike solutions. Assume that a uniform magnetic field $\mathbf{B}_0$ permeates a vertically stratified stationary plasma, with a uniform temperature (T_0) and with density and pressure distributions as given below:

$$\rho_0(z) = \text{const.} \times e^{-z/H} \tag{4.8}$$

$$p_0(z) = \text{const.} \times e^{-z/H} \tag{4.9}$$

satisfying the hydrostatic equilibrium (absence of any forces and motions)

$$-\frac{dp_0}{dz} - \rho_0 g = 0. \tag{4.10}$$

Here,

$$H = \frac{p_0}{\rho_0 g} \tag{4.11}$$

is the scale height, which depends on the medium in which one is interested. The linearized equations of motion are derived under the perturbations of the flow variables:

$$\rho = \rho_0 + \rho_1 ; \mathbf{v} = \mathbf{v}_1 ; p = p_0 + p_1 ; \mathbf{B} = \mathbf{B}_0 + \mathbf{B}_1,$$

where ρ, $\mathbf{v}$, p, and $\mathbf{B}$ are the density, velocity, pressure, and magnetic field, respectively. The linearized equations of motion (squares and products of the perturbed quantities are neglected) are as follows:

$$\frac{\partial \rho_1}{\partial t} + (\mathbf{v}_1 \cdot \nabla)\rho_0 + \rho_0(\nabla \cdot \mathbf{v}_1) = 0 \tag{4.12}$$

$$\rho_0 \frac{\partial \mathbf{v}_1}{\partial t} = -\nabla p_1 + (\nabla \times \mathbf{B}_1) \times \mathbf{B}_0/\mu - \rho_1 g \hat{\mathbf{z}} - 2\rho_0 \mathbf{\Omega} \times \mathbf{v}_1 \tag{4.13}$$

$$\frac{\partial p_1}{\partial t} + (\mathbf{v}_1 \cdot \nabla)p_0 - c_s^2 \left(\frac{\partial \rho_1}{\partial t} + (\mathbf{v}_1 \cdot \nabla)\rho_0 \right) = 0 \tag{4.14}$$

$$\frac{\partial \mathbf{B}_1}{\partial t} = \nabla \times (\mathbf{v}_1 \times \mathbf{B}_0) \tag{4.15}$$

$$\nabla \cdot \mathbf{B}_1 = 0, \tag{4.16}$$

where $c_s^2 = \gamma p_0/\rho_0$ is the sound speed. The advantage of linearizing the equations of motion is that by simple algebraic simplifications, the equations of motions involving the perturbed variables can be reduced to an equation with one variable only, say, for example, the perturbed velocity. Taking the time derivative of Eq. (4.13) and replacing the terms $\partial \rho_1/\partial t$, $\partial p_1/\partial t$, and $\partial \mathbf{B}_1/\partial t$ from Eqs. (4.12), (4.14), and (4.15), respectively, will result in a single equation as follows:

$$\frac{\partial^2 \mathbf{v}_1}{\partial t^2} = c_s^2 \nabla(\nabla \cdot \mathbf{v}_1) - (\gamma - 1)g\hat{\mathbf{z}}(\nabla \cdot \mathbf{v}_1) - g\nabla v_{1z} - 2\Omega \times \frac{\partial \mathbf{v}_1}{\partial t}$$

$$+ [\nabla \times (\nabla \times (\mathbf{v}_1 \times \mathbf{B}_0))] \times \frac{\mathbf{B}_0}{\mu \rho_0}. \tag{4.17}$$

Having derived a single equation for $\mathbf{v}_1$, we seek plane-wave solutions of the form

$$\mathbf{v}_1(\mathbf{r}, t) = \mathbf{v}_1 e^{i(\mathbf{k} \cdot \mathbf{r} - \omega t)}.$$

Here $\mathbf{k}$ is the wavenumber vector and ω the frequency. The period of the wave can be defined as $2\pi/\omega$, whereas the wavelength λ is just $2\pi/k$. The direction of propagation of the wave is given by $\hat{\mathbf{k}}(\equiv \mathbf{k}/k)$. By assuming plane-wave solutions, one replaces $\partial/\partial t$ by $-i\omega$ and ∇ by $i\mathbf{k}$ in Eq. (4.17).

For the special case when the magnetic field $\mathbf{B}_0 = 0$, Eq. (4.17) simplifies to

$$\omega^2 \mathbf{v}_1 = c_s^2 \mathbf{k}(\mathbf{k} \cdot \mathbf{v}_1) + i(\gamma - 1)g\hat{\mathbf{z}}(\mathbf{k} \cdot \mathbf{v}_1) + igk v_{1z} - 2i\omega\Omega \times \mathbf{v}_1. \tag{4.18}$$

It is important to derive the dispersion relation $\omega = \omega(\mathbf{k})$, which gives the frequency as a function of the wavenumber $\mathbf{k}$, being dependent on gravity and magnetic field. Equation (4.18) is a vector equation, with the three velocity components. Once the equation is written in terms of algebraic equations, the dispersion relation can be derived by putting the coefficients and setting the determinant to zero. The velocity $\mathbf{v}_{ph} = (\omega/k)\hat{\mathbf{k}}$ is known as the phase velocity of the wave. Its magnitude (ω/k) gives the speed of propagation in the direction $\hat{\mathbf{k}}$ for a wave specified by a single wavenumber. In most of the problems in wave theory, the dispersion relation among ω, the frequency, and wavenumber $\mathbf{k}$ is not necessarily a linear relationship. If the relationship is linear, then the waves are termed nondispersive. Otherwise, they are dispersive. For such waves, one can define the concept of a packet (or group) of waves with a range of wavenumbers, traveling with the group velocity $(\mathbf{V}_g)$, with components given by

$$V_{gx} = \frac{\partial \omega}{\partial k_x}; V_{gy} = \frac{\partial \omega}{\partial k_y}; V_{gz} = \frac{\partial \omega}{\partial k_z}.$$

The physical interpretation of the group velocity is that it is the velocity at which energy is transmitted and in general is different both in magnitude and in direction from the phase velocity of the wave.

4.2 Sound Waves

In order to study sound waves, we should set $g = \mathbf{B}_0 = \Omega = 0$; namely, the effects of gravity, magnetic field, and rotation will be ignored. This will imply that the only restoring force will be the pressure gradient. Equation (4.18) in this case will reduce to

$$\omega^2 \mathbf{v}_1 = c_s^2 \mathbf{k}(\mathbf{k} \cdot \mathbf{v}_1). \tag{4.19}$$

Taking the scalar product of the above equation with $\mathbf{k}$ and assuming that $\mathbf{k} \cdot \mathbf{v}_1 \neq 0$, we find

$$\omega^2 = k^2 c_s^2. \tag{4.20}$$

The above equation has two solutions. For the propagating disturbances, the dispersion relation for the sound (acoustic) waves is written as

$$\omega = k c_s. \tag{4.21}$$

It is interesting to note that sound waves propagate equally in all directions (isotropically) with a phase speed given by

$$v_{ph}\left(\equiv \frac{\omega}{k}\right) = c_s. \tag{4.22}$$

Differentiating the dispersion relation with respect to k gives the group velocity, which in this case is same as the phase velocity:

$$V_g\left(\equiv \frac{d\omega}{dk}\right) = c_s. \tag{4.23}$$

The linear theory of sound waves, in the absence of other restoring forces, is rather simple. However, in most cases, compressibility, which is responsible for the sound waves, gets coupled with other forces such as gravity, magnetic field, and so forth. Also, when there are no shocks (density and pressure discontinuities), these waves are longitudinal in the sense that the velocity perturbation $(\mathbf{v}_1)$ is in the direction of the propagation of $(\mathbf{k})$, the wavenumber.

4.3 Alfvén Waves

It is well known in physics that the tension in an elastic string allows transverse waves to propagate along the string. In a similar way, it is reasonable to assume that the magnetic tension will produce transverse waves that will propagate along a magnetic field $\mathbf{B_0}$ with a speed $[(B_0^2/(\mu\rho_0)]^{1/2}$. This expression is called the Alfvén speed and is given by

$$V_A = \frac{B_0}{(\mu\rho_0)^{1/2}}. \tag{4.24}$$

Also, it is well known that the pressure of a gas obeying the adiabatic law, $p/\rho^\gamma = $ constant, will produce (longitudinal) sound (acoustic) waves whose phase speed will be $(\gamma p_0/\rho_0)^{1/2}$. In a similar way, one might expect the magnetic pressuree $p_m = B_0^2/(2\mu)$ to generate longitudinal magnetic waves propagating across the magnetic field with a phase speed given by $[B_0^2/(\mu\rho_0)]^{1/2}$, which is the Alfvén speed.

The Lorentz force $\mathbf{j} \times \mathbf{B}$ can, in principle, drive a magnetic wave, which can propagate either along or across (may be oblique) the field. In what follows, we assume that the magnetic field dominates the equilibrium state, so that the restoring forces, such as pressure, gravity, and rotation, may be neglected. The wave equation for such a scenario may be written as

$$\omega^2 \mathbf{v}_1 = [\mathbf{k} \times (\mathbf{k} \times (\mathbf{v}_1 \times \hat{\mathbf{B}}_0))] \times \hat{\mathbf{B}}_0 V_A^2. \tag{4.25}$$

$\hat{\mathbf{B}}_0$ is the unit vector in the direction of the magnetic field.

The vector identity on the right-hand side of Eq. (4.25) may be simplified to yield

$$\omega^2 \mathbf{v}_1/V_A^2 = (\mathbf{k} \cdot \hat{\mathbf{B}}_0)^2 \mathbf{v}_1 - (\mathbf{k} \cdot \mathbf{v}_1)(\mathbf{k} \cdot \hat{\mathbf{B}}_0)\hat{\mathbf{B}}_0 + [(\mathbf{k} \cdot \mathbf{v}_1) - (\mathbf{k} \cdot \hat{\mathbf{B}}_0)(\hat{\mathbf{B}}_0 \cdot \mathbf{v}_1)]\mathbf{k}.$$

Assume that the propagation vector of the wave $\mathbf{k}$ makes an angle θ with the equilibrium magnetic field $\mathbf{B}_0$. Then the above equation can be rewritten as

$$\omega^2 \mathbf{v}_1/V_A^2 = k^2\cos^2\theta \mathbf{v}_1 - (\mathbf{k} \cdot \mathbf{v}_1)k\cos\theta \hat{\mathbf{B}}_0 + [(\mathbf{k} \cdot \mathbf{v}_1) - k\cos\theta(\hat{\mathbf{B}}_0 \cdot \mathbf{v}_1)]\mathbf{k}. \tag{4.26}$$

If we take scalar product of the above equation with $\hat{\mathbf{B}}_0$, we get

$$\hat{\mathbf{B}}_0 \cdot \mathbf{v}_1 = 0. \tag{4.27}$$

In a similar way, if we take the scalar product with respect to $\mathbf{k}$, we get

$$(\omega^2 - k^2 V_A^2)(\mathbf{k} \cdot \mathbf{v}_1) = 0, \tag{4.28}$$

which has two distinct solutions.

Let's turn our attention to a simple polarization of Alfvén waves. Consider the waves propagating along the z-axis. The straight and homogeneous magnetic field will be in the xz- plane and has two components:

$$\mathbf{B}_0 = B_0\sin\alpha\hat{x} + B_0\cos\alpha\hat{z},$$

where B_0 is the absolute value of the magnetic field, and α is the angle between the magnetic field and the z-axis. In the linear analysis, the set of MHD equations splits into two uncoupled subsets for the different components of the velocity and magnetic field. The first will be the Alfvén wave. The other is the magnetoacoustic waves, which will be discussed in detail later in the chapter. The situation when $\mathbf{B}_0 \parallel \hat{z}$ can lead to two linearized polarized plane Alfvén waves, one perturbing V_y, B_y and the other V_x, B_x (Nakariakov 2002).

For harmonic perturbations proportional to $\exp(i\omega t - ikz)$, where ω is the frequency and k the wavenumber, a combination of two linearly polarized waves will lead to elliptically polarized Alfvén waves given by

Fig. 4.1 The magnetic field perturbation in the xy-plane; from Nakariakov (2002)

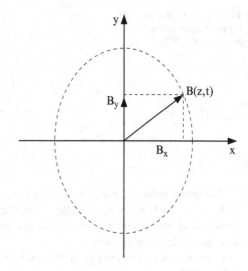

$$B_y = A\cos(\omega t - kz)$$
$$B_x = B\sin(\omega t - kz). \tag{4.29}$$

Here A and B are constants. The vector of the magnetic field perturbation rotates along an ellipse at the xy-plane. The special case when $A = B$ describes a circularly polarized case with $|\mathbf{B}|$ = constant. Figure 4.1 describes the polarization of the Alfvén waves. The circularly polarized Alfvén waves are an exact solution of the ideal MHD equations for a homogeneous medium.

4.4 Shear Alfvén Waves

In the incompressible limit wherein $(\nabla \cdot \mathbf{v}_1 = 0)$, we have

$$\mathbf{k} \cdot \mathbf{v}_1 = 0. \tag{4.30}$$

∇ is replaced by $\mathbf{k}$. Equation (4.26) is simplified, after taking the positive square root:

$$\omega = kV_A\cos\theta \tag{4.31}$$

for Alfvén waves, which sometimes is referred to as shear Alfvén waves. The positive square root describes waves propagating in the same direction as the magnetic field, while the negative root would describe waves propagating in the opposite direction. The variation of the phase speed with θ is presented as a polar diagram that has two circles of diameter V_A (see Fig. 4.2).

Fig. 4.2 The polar diagram
for the Alfvén waves;
from Priest (1982)

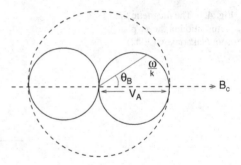

The group velocity, which is obtained by differentiating the dispersion relation with respect to k, gives $\mathbf{v}_g = V_A\hat{\mathbf{B}}_0$. Thus, the energy is transferred at the Alfvén speed along the magnetic field, although the individual waves can travel at a different angle of inclination.

In dealing with the energy of Alfvén waves, let us now return to the Lorentz force, which is written as

$$\mathbf{j}_1 \times \mathbf{B}_0 = (\mathbf{k} \times \mathbf{B}_1) \times \mathbf{B}_0\mu$$
$$= (\mathbf{k} \cdot \mathbf{B}_0)\mathbf{B}_1/\mu - (\mathbf{B}_0 \cdot \mathbf{B}_1)\mathbf{k}/\mu. \tag{4.32}$$

The first term on the right-hand side of the above equation represents the magnetic tension, while the second term is the magnetic pressure. In a sense, the driving force for the Alfvén waves is mostly the magnetic tension alone. The ratio of the magnetic energy to the kinetic energy is given by

$$\left[B_0^2/(2\mu)\right] / [(1/2)\rho_0 v_1^2],$$

which is almost equal to 1. Thus, the Alfvén waves allow an equipartition between the magnetic and kinetic energies.

4.5   Compressional Alfvén Waves

One of the solutions of Eq. (4.28) is

$$\omega = kV_A, \tag{4.33}$$

which is known as a compressional Alfvén wave. In this case, the phase speed is always V_A regardless of the angle of propagation. The group velocity is given by $\mathbf{V}_g = V_A\mathbf{k}$, so that the energy is propagated isotropically.

The velocity perturbation $\mathbf{v}_1$ is in the direction normal to $\mathbf{B}_0$ and lies in the plane $(\mathbf{k}, \mathbf{B}_0)$. Thus, it possesses components along and transverse to $\mathbf{k}$ in general, which will result in changes in both density and pressure. For the special case when $\theta = 0$ (4.31), the compressional Alfvén wave is transverse and is identical with the ordinary Alfvén wave. In this case, it is completely dominated by magnetic tension and does not produce any compression. The compressional wave reaches the incompressible limit when the angle of propagation is zero, that is, along the magnetic field.

Observational signatures: In the last decade, the progress in the spatial and time resolution of instruments brought up abundant evidence of the MHD wave activity in the solar corona. Hinode (a Japanese word meaning sunrise) was launched by the Japanese Space Agency in September 2006, joining a number of other outstanding high-resolution contemporary space missions, such as Yohkoh, the Solar and Heliospheric Observatory (SOHO), the Transition Region and Coronal Explorer (TRACE), the Reuven Ramaty High Energy Solar Spectroscopic Imager (RHESSI), and the Solar Terrestrial Relations Observatory (STEREO), all of which have been used to study the important problem of heating of the solar corona, the acceleration of the solar wind, and plasma particles. Hinode had three high-resolution instruments: the Solar Optical Telescope (SOT); the Extreme Ultraviolet Imaging Spectrometer (EIS); and the X-Ray Telescope (XRT), which has revealed some of the secrets of the corona. Among the many theories on the heating of the solar corona, the most compelling one involves the Alfvén waves, predicted theoretically by Hannes Alfvén. It is well known that in a uniform plasma, there are three distinct types of MHD waves: slow and fast magnetoacoustic waves, and Alfvén waves. The first two types of waves have an acoustic character, modified by the magnetic field, whereas the Alfvén waves exist purely because of the presence of a magnetic field. There are many examples of the observation of slow and fast MHD waves in the solar atmosphere; however, their energy does not seem to be near enough to explain the coronal heating. On the other hand, among MHD wave theorists, the Alfvén waves are considered the most promising energy transporter. It has always been expected that once these waves are generated, they will easily propagate along the magnetic flux tubes, the building block of the solar atmosphere, or along magnetic field lines at constant magnetic surfaces.

According to some observations using Hinode's XRT and SOT instruments, there is some evidence on the signatures of Alfvén waves. The estimation of the energy density of the observed waves has shown that they are sufficient to accelerate the solar wind and heat the solar corona. Also, the existence of waves has been suggested by observing the corona using the Coronal Multi-Channel Polarimeter (CoMP). The revelation that the waves are omnipresent in the corona is an important development. However, many uncertainties (such as instrumental sources of systematic error, uncertainties on the field inclination and atomic polarization, 3D geometry) will have to be resolved in the future, and it is very clear that the discovery has major implication for coronal physics, for instance, in the context of coronal heating and coronal seismology. The observed oscillations only had a significant contribution in the Fourier power spectrum of the velocity signal. No oscillations were observed in the intensity or line width. The interpretation that the observations amounted to Alfvén waves was based on the facts that the observed phase speeds are much larger than the sound speed, the waves propagate along the field lines, and the waves are seen to be incompressible. However, some theorists have interpreted these waves as kink waves and not Alfvén waves.

The properties of the plume and interplume regions in a polar coronal hole and the role of waves in the acceleration of the solar wind have been studied recently (O'Shea et al. 2007; Banerjee et al. 2009), to detect whether Alfvén waves are present in the polar coronal holes through variations in EUV line widths. Using spectral observations performed over a polar coronal hole region with the EIS spectrometer on Hinode, the variation in the line width and electron density as a function of height, with the density-sensitive line pairs of Fe XII $186.88\,\mathring{A}$ and $195.119\,\mathring{A}$ and Fe XIII $203.82\,\mathring{A}$ and $202.04\,\mathring{A}$, was used. For the polar region, the line width data showed that the nonthermal line-of-sight velocity increases from $26\,km^{-1}$ at 10" above the limb to $42\,km^{-1}$ some 150" (i.e., $110,000\,km$) above the limb. The electron density shows a decrease from $3.3 \times 10^9\,cm^{-3}$ to $1.9 \times 10^8\,cm^{-3}$ over the same distance. These results imply that the nonthermal velocity is inversely proportional to the quadratic root of the electron density, in excellent agreement with what is predicted for undamped radially propagating linear Alfvén waves. The data provide signatures of Alfvén waves in the polar coronal hole regions, which could be important for the acceleration of the solar wind.

A 2010 study on accelerating disturbances in the polar plume and interplume Gupta et al. (2010) presents EIS/Hinode & SUMER/SoHO joint observations, allowing the first spectroscopic detection of accelerating disturbances as recorded with coronal lines in interplume and plume regions of a polar coronal hole. With the help of time-distance radiance maps, the presence of propagating disturbances in a polar interplume region with a period of 15–20 min and a propagation speed increasing from $130 \pm 14\,km/s$ just above the limb, to $330 \pm 140\,km/s$ around $160\,km/s$ above the limb, was detected. These disturbances can also be traced to originate from a bright region of the on-disk part of the coronal hole, where the propagation speed was found to be in the range of 25 ± 1.3 to $38 \pm 4.5\,km/s$, with the same periodicity. These on-disk bright regions can be visualized as the base of the coronal funnels. The adjacent plume region also shows the presence of a propagating disturbance with the same range of period but with propagation speeds in the range of 135 ± 18 to $165 \pm 43\,km/s$ only. A comparison between the time-distance radiance map of both regions indicates that the disturbances within the plumes are not observable (may be getting dissipated) far off-limb, whereas this is not the case in the interplume region.

Nonlinear studies: A possible role of MHD waves in accelerating the high-speed solar wind and the heating of the plasma in the open magnetic structures of the solar corona has been discussed in the literature. The presence of MHD waves in the open structures of the solar corona is more or less well established. The weakly nonlinear dynamics of linearly polarized, spherical Alfvén waves in coronal holes has been investigated (Nakariakov et al. 2000). An evolutionary equation, combining the effects of spherical stratification, nonlinear steepening, and dissipation due to shear viscosity, has been derived. The evolution equation is similar to the scalar Cohen–Kulsrud–Burges equation given by

$$\frac{\partial V_\phi}{\partial R} - \frac{1}{4HR^2}V_\phi - \frac{1}{4V_A(V_A^2 - c^2 s)}\frac{\partial V_\phi^3}{\partial \tau} - \frac{\hat{v}}{2V_A^3}\frac{\partial^2 V_\phi}{\partial \tau^2} = 0.$$

Here, both the Alfvén and sound speeds are measured in units of the Alfvén speed at the base of the corona. The scale height H is measured in units of the solar radius, while the normalized viscosity $\hat{v} = v/[R_\odot V_A(R_\odot)]$. V_ϕ is the transverse component of the velocity, $\tau = t - \int \frac{d\tau}{V_A}$, $R = \varepsilon \tau$, and ε is a small, positive parameter of the order of the nonlinearity.

The scenario of nonlinear dissipation is independent of viscosity. The dissipation rate is stronger for the highest amplitudes and depends weakly on the wave period and the temperature of the atmosphere. The nonlinear distortion of the wave shape is accompanied by the generation of longitudinal motions and density perturbations.

The importance of nonlinear fast magnetosonic waves in solar coronal holes has been developed by assuming the coronal hole as a slab of cold plasma threaded by a vertical, uniform magnetic field. A periodic driver acting at the coronal base drives the velocity component normal to the equilibrium magnetic field. The nonlinear terms in the MHD equations give rise to excitation of the velocity component parallel to the equilibrium **B**, with a lower amplitude than the normal component. The nature of the nonlinear interactions in the MHD equations determines the frequency of these modes. They are quadratic in the case of the parallel component, while for the normal component, they are cubic in nature.

The effect of steady flows (low-speed) directed along the magnetic field on the nonlinear coupling of MHD waves has been studied. The effect is similar to the Alfvén wave phase mixing in a static, inhomogeneous medium, which leads to the production of steep transverse gradients in the plasma parameters, which increases the dissipation. The transverse gradients in the total pressure, produced by phase mixing, lead to the secular generation of obliquely propagating fast magnetosonic waves, at double the frequency and wavenumber of the source Alfvén waves. The secular growth of density perturbations, connected with fast waves, takes place for flow speeds that are considerably below the thresholds of the Kelvin–Helmholtz and negative-energy wave instabilities.

The steady state of nonlinear, small-amplitude, quasi-resonant Alfvénic oscillations in a homogeneous dissipative hydromagnetic cavity, forced by shear motion, has been studied Nocera and Ruderman (1998). They showed that in the case of strong nonlinearity, these oscillations can be represented, to a leading order, by a sum of two solutions in the form of oppositely propagating waves. The resulting evolution equation (nonlinear) admits multiple solutions that depend on the Reynolds number, Re, and $\triangle$, which is the tuning between the frequency of the boundary forcing and the first Alfvén eigenmode of the cavity. The purpose of this study is to explain certain bright events in the solar atmosphere.

The nonlinear viscous damping of surface Alfvén waves in polar coronal holes, using data on electron density, temperature, and magnetic field near the edges, has been studied Narain and Sharma (1998). They found that in the nonlinear regime, the viscous damping of surface Alfvén waves becomes a visible mechanism for solar coronal plasma heating, when the magnetic field is sufficiently stronger.

4.6 Magnetoacoustic Waves

The name itself suggests that these waves are generated when there is a coupling between the magnetic field and pressure fluctuations. The starting point for a discussion of these waves will be Eq. (4.18), with the magnetic field included, while g and Ω are not present. By simple algebra, one can write the equation as

$$\omega^2 \mathbf{v}_1 / V_A^2 = k^2 \cos^2\theta \mathbf{v}_1 - (\mathbf{k} \cdot \mathbf{v}_1)k\cos\theta \hat{\mathbf{B}}_0$$

$$+[(1 + c_s^2/V_A^2)(\mathbf{k} \cdot \mathbf{v}_1) - k\cos\theta(\hat{\mathbf{B}}_0 \cdot \mathbf{v}_1)]\mathbf{k}. \tag{4.34}$$

Taking the scalar product of $\mathbf{v}_1$ with $\mathbf{k}$ and $\mathbf{B}_0$, we have

$$(-\omega^2 + k^2 c_s^2 + K^2 V_A^2)(\mathbf{k} \cdot \mathbf{v}_1) = k^3 V_A^2 \cos\theta(\hat{\mathbf{B}}_0 \cdot \mathbf{v}_1) \tag{4.35}$$

and

$$k\cos\theta c_s^2(\mathbf{k} \cdot \mathbf{v}_1) = \omega^2(\hat{\mathbf{B}}_0 \cdot \mathbf{v}_1). \tag{4.36}$$

Eliminating $(\mathbf{k} \cdot \mathbf{v}_1)/(\hat{\mathbf{B}}_0 \cdot \mathbf{v}_1)$ between the above two equations leads to the dispersion relation for magnetoacoustic (or magnetosonic) waves:

$$\omega^4 - \omega^2 k^2 (c_s^2 + V_A^2) + c_s^2 V_A^2 k^4 \cos^2\theta = 0. \tag{4.37}$$

If the waves are outward-propagating, where $(\omega/k > 0)$, then there are two distinct solutions, as given by

$$\omega/k = \left[(1/2)(c_s^2 + V_A^2) \pm (1/2)\sqrt{c_s^4 + V_A^4 - 2c_s^2 V_A^2 \cos\theta}\right]^{1/2}. \tag{4.38}$$

One of the frequencies will be higher than the other. The higher-frequency mode is generally termed the fast magnetoacoustic wave, while the lower-frequency mode will be called the slow magnetoacoustic wave. The phase speed of the Alfvén wave lies in between the phase speed of the slow and fast modes. Thus, it is referred to in the literature as the intermediate mode.

It is very clear from the expression of the phase speed of the two magnetoacoustic modes that they depend on the direction of propagation. For the case when the propagation is along the magnetic field, the phase speed ω/k is either c_s or V_A, while for propagation across the field ($\theta = \pi/2$), ω/k is $(c_s^2 + V_A^2)^{1/2}$ or 0. As the angle approaches $\pi/2$, the phase speed of the wave tends to

$$c_T = \frac{V_A c_s}{(V_A^2 + c_s^2)^{1/2}}$$

Fig. 4.3 The polar diagram for the magnetoacoustic waves; from Nakariakov (2002)

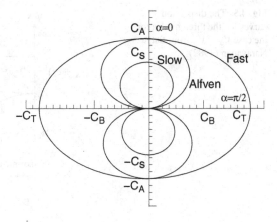

Fig. 4.4 The group velocity for magnetoacoustic waves; from Nakariakov (2002)

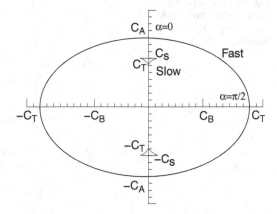

c_T, which is the component of the phase velocity along the field for propagation almost perpendicular to the field. The wavelength along the field will be larger than the wavelength across the field. c_T is also referred to as the cusp speed for the group velocity of the slow wave. The polar diagram for the magnetoacoustic waves and their group velocity are given in Figs. 4.3 and 4.4, respectively.

The magnetoacoustic waves that we have discussed so far may be considered to be a sound wave, modified due to the presence of a magnetic field, and the compressional Alfvén wave, modified by the pressure. The limiting cases of vanishing magnetic field and gas pressure are interesting as in these cases the slow mode disappears, while the fast mode becomes the ordinary sound wave and a compressional Alfvén wave, respectively.

When the plasma $\beta = 2\mu p_0/B_0^2$, the ratio of the magnetic pressure to the gas pressure is larger than unity, then the ratio of the sound speed to the Alfvén speed

Fig. 4.5 The dispersion
curves for the three modes for
the case $V_A < V_s$; from Satya
Narayanan (1982)

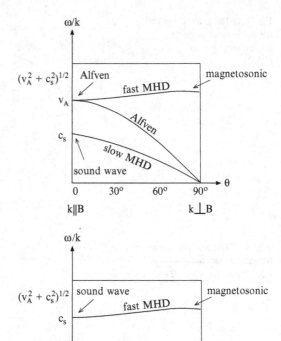

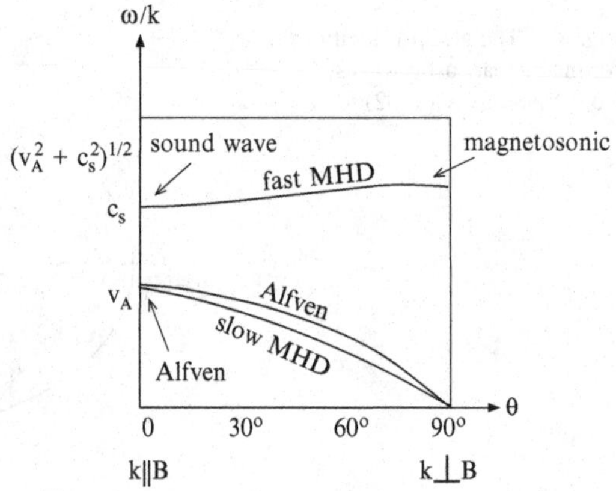

Fig. 4.6 The dispersion
curves for the three modes for
the case $V_A > V_s$; from Satya
Narayanan (1982)

given by c_s^2/V_A^2 is also much larger than 1. The dispersion relation for the fast and
slow magnetoacoustic modes reduces to

$$\omega/k \approx c_s$$

and

$$\omega/k \approx V_A \cos\theta, \tag{4.39}$$

respectively. For the slow mode,

$$\mathbf{k} \cdot \mathbf{v} \approx \frac{V_A}{c_s^2} \cos\theta (\mathbf{B}_0 \cdot \mathbf{v}_1),$$

which is much less than unity, so that the disturbance is more or less incompressible.
The polar diagram for the group velocity of these modes implies that the slow-mode
energy is seen to propagate in a narrow cone about the magnetic field, while the fast
mode has energy that propagates isotropically. A schematic picture of the dispersion
curves for the Alfvén, slow, and fast modes is given in Figs. 4.5 and 4.6.

Table 4.1 MHD waves

Wave type	Behavior	Weak field	Strong field
Fast	Isotropic $v_{\text{ph}} \sim \max(v_s, V_A)$	Gas pressure $\mathbf{v} \Vert \mathbf{k}$	Magnetic pressure $\mathbf{v} \perp \mathbf{B}_0$
Slow	Propagates approximately along $\mathbf{B}_0$ $v_{\text{ph}} \sim \min(v_s, V_A)$	Magnetic tension $\mathbf{v} \perp \mathbf{k}$	Gas pressure $\mathbf{v} \Vert \mathbf{B}_0$
Alfvén	Propagates along $\mathbf{B}_0$ $v_{\text{ph}} = V_A$	Magnetic tension $\mathbf{v} \perp \mathbf{k}$ and $\mathbf{B}_0$	

The above table compares the properties of the Alfvén and magnetoacoustic (fast and slow) waves (Table 4.1).

4.7 Internal and Magnetoacoustic Gravity Waves

In studying the properties of internal gravity waves, one assumes that a parcel of fluid, displaced vertically over a distance from the equilibrium, remains in pressure equilibrium with its surroundings and that the density changes inside the fluid parcel are adiabatic (no heat transfer). The first assumption is valid if the motion is slow compared to the sound waves, and the second holds if the motion is fast enough for the entropy to be preserved. The equilibrium is achieved when the pressure gradient is balanced by gravity; that is,

$$\frac{dp_0}{dz} = -\rho_0 g. \tag{4.40}$$

Outside the fluid parcel, the pressure and density at a height $z + \delta z$ will be given by $p_0 + \delta p_0$ and $\rho_0 + \delta \rho_0$, so that the above equation may be rewritten as

$$\delta p_0 = -\rho g \delta z \tag{4.41}$$

and

$$\delta \rho_0 = -\frac{d\rho_0}{dz} \delta z. \tag{4.42}$$

Since we have assumed that the pressure and density obey the adiabatic relationship, namely, $p/\rho^\gamma = \text{constant}$, the change in the pressure will be related to the sound speed in the medium; that is, $\delta p = c_s^2 \delta \rho$. Replacing the expression for δp into Eq. (4.41) gives an expression for the internal density change as

$$\delta \rho = -\frac{\rho_0 g \delta z}{c_s^2}. \tag{4.43}$$

By simple algebra, one can show that

$$g(\delta\rho_0 - \delta\rho) = -N^2\rho_0\delta z. \tag{4.44}$$

The expression

$$N^2 = -g\left(\frac{1}{\rho_0}\frac{d\rho_0}{dz} + \frac{g}{c_s^2}\right) \tag{4.45}$$

when N is real is known as the Brunt–Vaisala frequency. It plays an important role in studying the static stability of stratified flows. For static stability, $N^2 > 0$.

In the presence of a horizontal magnetic field, the Brunt–Vaisala frequency is modified to

$$N^2 = -g\left(\frac{1}{\rho_0}\frac{d\rho_0}{dz} + \frac{g}{c_s^2 + V_A^2}\right).$$

In the case of uniform temperature (isothermal), N becomes

$$N^2 = \frac{g^2}{c_s^2}\left(\gamma - \frac{c_s^2}{c_s^2 + V_A^2}\right).$$

The condition $N^2 > 0$ is usually referred to as the Schwarzschild criterion for convective stability.

In order to derive the dispersion relation for the internal gravity waves, the starting point will be Eq. (4.18), with the term Ω absent. Taking the scalar product with $\mathbf{k}$ and $\hat{\mathbf{z}}$ in turn and gathering together the terms in v_{1z} and $\mathbf{k}\cdot\mathbf{v}_1$, we obtain

$$igk^2v_{1z} = (\omega^2 - c_s^2k^2 + i(\gamma-1)gk_z)(\mathbf{k}\cdot\mathbf{v}_1)$$
$$(\omega^2 - igk_z)v_{1z} = (c_s^2k_z + i(\gamma-1)g)(\mathbf{k}\cdot\mathbf{v}_1). \tag{4.46}$$

Eliminating $(\mathbf{k}\cdot\mathbf{v}_1)/v_{1z}$ between the above expressions yields

$$(\omega^2 - igk_z)(\omega^2 - c_s^2k^2 + i(\gamma-)gk_z) = igk^2(c_s^2k_z + i(\gamma-1)g). \tag{4.47}$$

The frequency of waves we will be interested in should be of the order of the Brunt–Vaisala frequency and much smaller than the frequency of sound waves, which implies that

$$\omega \approx \frac{g}{c_s} \ll kc_s.$$

Applying the above condition to Eq. (4.47), we get

$$\omega^2c_s^2 \approx (\gamma-1)g^2(1 - k_z^2/k^2). \tag{4.48}$$

In terms of the inclination angle $\theta = \cos^{-1}(k_z/k)$ and N, the expression can be simplified to yield

$$\omega = N\sin\theta. \tag{4.49}$$

The above expression is the relation for internal gravity waves. This is different from surface gravity waves, which propagate along the surface between two fluids.

Acoustic and gravity modes occur when both compressibility and buoyancy forces are present. In general, they remain as distinct modes, but are modified under special circumstances. Taking the scalar product of the basic equation (4.18) with $\mathbf{k}$ and $\hat{\mathbf{z}}$ in turn yields Eq. (4.46), which in turn gives the dispersion relation

$$\omega^2(\omega^2 - N_s^2) = (\omega^2 - N^2\sin^2\theta_g)k'^2c_s^2, \tag{4.50}$$

where

$$N_s = \frac{\gamma g}{2c_s}\left(\frac{c_s}{2H}\right)$$

$$N = \frac{(\gamma-1)^{1/2}g}{c_s}$$

$$\sin^2\theta_g = 1 - \frac{k_z^2}{k^2}.$$

$$\mathbf{k}' = \mathbf{k} + i\frac{\gamma g}{2c_s^2}\hat{\mathbf{z}}$$

θ_g in Eq. (4.50) is the angle between the vector $\mathbf{k}'$ and the vertical. $N \geq N_s$ in general and is equal when $\gamma = 2$.

In the special case when $\omega \ll k'c_s$, the dispersion relation (4.50) reduces to $\omega = N\sin\theta_g$ for gravity waves, whereas for $\omega \gg N$, it becomes $\omega = k'c_s$, pure acoustic waves. For vertical propagation, the wave exists only if $\omega > N_s$ (4.50) can be rewritten as follows:

$$k_z'^2\omega^2c_s^2 = \omega^2(\omega^2 - N_s^2) - (\omega^2 - N^2)k_x^2c_s^2.$$

For the case when ω^2 and k_x^2 are positive, there are vertically propagating waves ($k_z'^2 > 0$) provided

$$\omega^2(\omega - n_s^2) > (\omega^2 - N^2)k_x^2c_s^2.$$

The effect of the magnetic field on the acoustic-gravity waves complicates the situation with an extra restoring force and preferred direction in addition to the force due to gravity. The Alfvén wave propagates unaltered; however, the magnetic field modifies the propagation characteristics of the acoustic-gravity waves to give rise to two magnetoacoustic-gravity waves.

When the Coriolis force is absent, the equation for the wave propagation in a fluid with a uniform magnetic field, temperature, and a density proportional to $\exp(-z/H)$ allows the solution

$$\omega^2 = k^2V_A^2\cos^2\theta$$

in addition to the two magnetoacoustic-gravity modes. In the presence of a magnetic field, plane-wave solutions exist when $(kH)^{-1} \ll 1$, namely, if

$$N_s \ll kc_s,$$

which means the wavelength is very much shorter than the scale height H. Using the above expression and Eq. (4.18), the dispersion relation for the magnetoacoustic-gravity waves can be obtained by eliminating $\mathbf{v}_1$ and using the above expression to yield

$$\omega^4 - \omega^2 k^2 (c_s^2 + V_A^2) + k^2 c_s^2 N^2 \sin^2 \theta_g + k^4 c_s^2 V_A^2 \cos^2 \theta = 0, \qquad (4.51)$$

where θ is the angle between $\mathbf{k}$ and the magnetic field. Interesting limiting cases arise. First, when N and V_A are neglected, then the acoustic waves are recovered, while for $V_A = 0$ and $\omega \ll kc_s$, the result for internal gravity is obtained. However, a point of caution is that for vanishing of the Alfvén speed, the full dispersion relation for acoustic-gravity waves is not recovered, for the simple reason that there is a coupling between the acoustic and gravity waves.

We shall briefly discuss the derivation of the wave equation for internal Alfvén gravity waves in stratified shear flows. The wave equation for the motion of a perfectly conducting fluid in the presence of a magnetic field with vertical density stratification will be discussed under the assumption that the motion is two-dimensional, variations being in the x- and z-directions. The fluid is inviscid, perfectly conducting, and adiabatic (Rudraiah and Venkatachalappa 1972; Satya Narayanan 1982). The Boussinesq approximation (a variation of the vertical coordinate except that the buoyancy term is neglected). The rotation is neglected. The basic shear is assumed to be $(U(z), 0)$. The perturbation is for the velocity components (v_x, v_z) and the magnetic field (B_x, B_z) with a uniform magnetic field $(B_0, 0)$, where B_0 is a constant. We also assume that

$$|v_x \partial/\partial x + v_z \partial/\partial z| \ll |\partial/\partial t + U \partial/\partial x| \qquad (4.52)$$

$$|B_x \partial/\partial x + B_z \partial/\partial z| \ll |\partial/\partial t + B_0 \partial/\partial x|. \qquad (4.53)$$

The linearized equations of motion can be reduced to a single equation for the vertical component of velocity:

$$(\partial/\partial t + U \partial/\partial x)^4 ((v_z)_{xx} + (v_z)_{zz}) - (\partial/\partial t + U \partial/\partial x)^3 (U_{zz}(v_z)_x)$$

$$+ (\partial/\partial t + U \partial/\partial x)^2 [N^2 (v_z)_{xx} - V_A^2 ((v_z)_{xxxx} + (v_z)_{xxzz})$$

$$+ V_A^2 (\partial/\partial t + U \partial/\partial x)[2U_z(v_z)_{xxxz} + U_{zz}(v_z)_{xxx}] - 2V_A^2 U_z^2 (v_z)_{xxxx} = 0. \qquad (4.54)$$

Here the subscripts x and z denote partial derivatives. Assuming a sinusoidal disturbance of the form

$$v_z \exp[i(kx - \omega t)],$$

the above equation reduces to a second-order differential equation for the vertical component of the velocity field:

$$\frac{d^2v_z}{dz^2} - (2k\Omega_A^2 U_z/\Omega_d(\Omega_d^2-\Omega_A^2))\frac{dv_z}{dz} + [k^2N^2/(\Omega_d^2-\Omega_A^2)+kU_{zz}/\Omega_d-k^2 \quad (4.55)$$

$$-2k^2\Omega^2 AU_z^2/\Omega_d^2(\Omega_d^2-\Omega_A^2)]v_z = 0. \quad (4.56)$$

Here $\Omega_d = \omega - kU$ is the Doppler-shifted frequency, and $\Omega_A = kV_A$ is the Alfvén frequency, with V_A the Alfvén velocity.

The above equation reduces to the famous Taylor–Goldstein equation when the magnetic field is neglected; that is,

$$\frac{d^2v_z}{dz^2} + [N^2/(U-c)^2 - U_{zz}/(U-c) - k^2]v_z = 0. \quad (4.57)$$

U is the basic shear. $c = \omega/k$ is the phase velocity of the wave. The wave equation for the internal Alfvén gravity waves in stratified shear flows is singular at $\Omega_d = 0, \pm\Omega_A$; namely, there are two magnetic singularities in addition to the hydrodynamic singularity. The effects of viscosity, thermal, and ohmic dissipation may intervene and prevent such singularities.

The propagation and dissipation of Alfvén gravity waves, considering viscosity as a damping mechanism, leads to a general dispersion relation for Alfvén gravity waves under WKB and Boussinesq approximation (Pandey and Dwivedi 2006):

$$Pk_z^4 + Qk_z^3 + Rk_z^2 + Sk_z + T = 0, \quad (4.58)$$

where

$$P = iV_A^2\omega^2\left(\omega - \frac{i\eta_0 k_\perp^2}{\rho_0}\right)$$

$$Q = \frac{i\eta_0 N^2 V_A^2\omega^2 k_\perp^2}{3\rho_0 g}$$

$$R = i\omega^3\left(V_A^2 k_\perp^2 - \omega^2\right) + \frac{\eta_0\omega^2 k_\perp^2}{\rho_0}\left(V_A^2 k_\perp^2 - 3\omega^2\right)$$

$$S = \frac{\eta_0 i N^2\omega^2 k_\perp^2}{3\rho_0 g}\left(V_A^2 k_\perp^2 - 3\omega^2\right)$$

and

$$T = i\omega^3 k_\perp^2(N^2 - \omega^2).$$

In the absence of viscosity, the above relation reduces to

$$V_A^2 k_z^4 + (V_A^2 k_\perp^2 - \omega^2)k_z^2 + k_\perp^2(N^2 - \omega^2) = 0. \quad (4.59)$$

Here k_z and $k_\perp$ are the wavenumbers in the z-direction and normal to the xz-plane. When viscosity is included, the wave normal surfaces do not differ significantly from the results obtained in an ideal MHD. From the dissipation point of view, for a wave frequency less than the Brunt–Vaisala frequency (i.e., $\omega < 1$), there are two types of damping: One is strong damping when $L_D \sim \lambda$ (L_D is the damping length), and the other is weak damping when $L_D \gg \lambda$. For wave frequencies greater than the Brunt–Vaisala frequency (i.e., $\omega > 1$), however, the oscillations are weakly damped. Thus, the Brunt–Vaisala frequency separates regimes of wave frequency in which damping is weak or strong.

Recently, the signature of gravity waves in the quiet solar atmosphere from TRACE 1,600 Å continuum observations were made with a 6-h time sequence of ultraviolet images. Fifteen uv bright points, 15 uv network elements, and 15 uv background regions in a quiet region were selected for the detailed analysis. The cumulative intensity values of these features were derived (Kariyappa et al. 2008). It was found that the uv bright points, the uv network elements, and uv background regions exhibit longer periods of intensity oscillations, namely, 5.5 h, 4.6 h, and 3.4 h, respectively, in addition to the small-scale intensity fluctuations. The longer periods of oscillations are related to solar atmospheric g-modes.

4.8 Phase Mixing and Resonant Absorption of Waves

The photospheric foot points of the magnetic field lines oscillate with a fixed frequency. Such a phenomenon leads to phase mixing (Heyvaerts and Priest 1983; Ireland and Priest 1997; Walsh 1999). Since each field line has its own Alfvén speed when the atmosphere is inhomogeneous (as is the situation in the Sun), the wave propagates at different phase speeds and moves out of phase. See the figure that sketches the phase mixing of Alfvén waves (see Fig. 4.7). This leads to large spatial

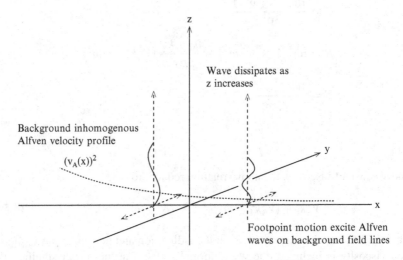

Fig. 4.7 Phase mixing of Alfvén waves; from Ireland and Priest (1997)

gradients until dissipation smooths them out, which will in turn extract energy. Wherever dissipation is important, this will lead to the heating of the field line. An important point is that this heat source will depend on time. Moreover, the frequency will vary with position and the amount of heat deposited.

A variation in the Alfvén speed would imply that the Alfvén waves have different wavelengths on the neighboring magnetic field lines. Thus, as the wave progresses in the z-direction, waves on the neighboring field lines move out of phase relative to each other. Consider the wave equation

$$\frac{\partial^2 v}{\partial t^2} = V_A^2(x)\frac{\partial^2 v}{\partial z^2},\tag{4.60}$$

which is the basic equation for the phase-mixing process in an inhomogeneous nondissipative system. If we assume the solution of the above equation is

$$v \approx \exp[-i(\omega t + k(x)z)],$$

where ω is the frequency and $k(x) = \omega/V_A(x)$, it implies that

$$\frac{\partial v}{\partial x} \sim vz\frac{dk}{dx}$$

such that the inclusion of a nonuniform background Alfvén velocity creates gradients in the x-direction that increase with z. Also, with dk/dx, sharper gradients may appear at lower heights depending on the inhomogeneity of the plasma.

For a dissipative system, assume that the initial vertical magnetic field (with $\mathbf{B}_0 = B_0(x)\hat{\mathbf{z}}$ and $\rho_0 = \rho_0(x)$) with the foot points oscillating with a frequency ω. This will introduce perturbations in the magnetic field and velocity as $b(x,z,t)$ and $v(x,z,t)$, respectively. The linearized equation of motion, together with the induction equation, will result in

$$\rho_0\frac{\partial v}{\partial t} = \frac{B_0}{\mu}\frac{\partial b}{\partial z} + \rho_0 v_v\nabla^2 v\tag{4.61}$$

$$\frac{\partial b}{\partial t} = B_0\frac{\partial v}{\partial z} + v_m\nabla^2 b,\tag{4.62}$$

where v_m is the magnetic diffusivity and v_v is the kinematic viscosity. The above equations can be combined to yield

$$\frac{\partial^2 v}{\partial t^2} = V_A^2(x)\frac{\partial^2 v}{\partial z^2} + (v_m + v_v)\left(\frac{\partial^2}{\partial x^2} + \frac{\partial^2}{\partial z^2}\right)\frac{\partial v}{\partial t}.\tag{4.63}$$

The term $V_A^2(x)\partial^2 v/\partial z^2$ takes different values on different field lines, which in turn will generate large horizontal gradients. However, these gradients will get smoothed out by the damping term $\partial^3 v/\partial t\partial z^2$. The damping term is not important for phase mixing and will be neglected. We now assume a solution of the above wave equation in the form

$$v \sim \hat{v}(x,z)\exp[-i(\omega t + k(x)z)].$$

We assume that for weak damping

$$\frac{1}{k}\frac{\partial}{\partial z} \ll 1, \tag{4.64}$$

and for strong phase mixing

$$\frac{z}{k}\frac{\partial k}{\partial x} \gg 1. \tag{4.65}$$

The solution for the equation (wave) can be written as

$$\hat{v}(x,z) = \hat{v}(x,0)\exp\left[-\frac{1}{6}(k(x)z/R^{1/3})^3\right], \tag{4.66}$$

where

$$R = \frac{\omega}{v_{\mathrm{m}} + v_{\mathrm{v}}}\left(\frac{\mathrm{d}}{\mathrm{d}x}\log k(x)\right)^2.$$

We briefly present one of the studies pertaining to phase-mixed shear Alfvén waves that was used to explain coronal heating (Heyvaerts and Priest 1983). The physical processes that occur when a shear Alfvén wave propagates in a structure with a large gradient of the Alfvén velocity was considered. These waves did not possess local resonances, but they undergo intense phase mixing during which the oscillations of neighboring field lines become rapidly out of phase. The resulting large growth of gradients dramatically enhances the viscous and ohmic dissipation. Both the cases of propagating and standing waves were studied. It was shown that after a sufficient time, phase mixing actually ensures the dissipation of all the wave mechanical energy the loop can pick up from the excitation. Instabilities developed in the phase-mixed flow play a decisive role in enhancing the dissipation by promoting the momentum exchange within the neighboring layers, which vibrate out of phase. One important conclusion is that the standing waves suffer tearing near the velocity nodes, while propagating waves appear to be stable. The phase-mixing process is able to ensure that the dissipation of shear Alfvén waves be in a permanent state of KH instability and tearing turbulence (Browning and Priest 1984). The above study has also been modified to include a stratified atmosphere in which the density decreases with height (statically stable situation). For a stratified atmosphere, the perturbed magnetic field and velocity behave quite differently depending on whether resistivity or viscosity is considered. Ohmic heating is spread out over a greater height range in a stratified medium, whereas viscous heating is not strongly influenced by the stratification.

The phase mixing of Alfvén waves in planar two-dimensional, open magnetic plasma configurations has been studied with the assumption that the characteristic vertical spatial scale is much larger than the horizontal scale and that the latter is of the order of a wavelength. The governing equation is derived using the WKB method, which looks similar to the diffusion equation, with the diffusion coefficient being dependent on spatial scales. Three different cases are discussed. In all three cases, at low heights, phase-mixed Alfvén waves damp at the same rate as in a

Fig. 4.8 Resonant absorption of Alfvén waves; from Walsh (1999)

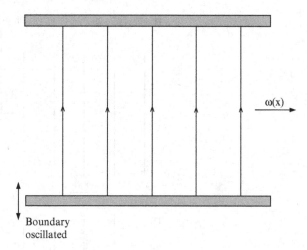

one-dimensional configuration. However, in the first and third cases, phase mixing operates only at low and intermediate heights and practically stops at heights larger than a few characteristic vertical length scales. The rate of damping of the energy flux with height due to phase mixing in two-dimensional configurations thus depends strongly on the particular form of the configuration.

The nonlinear excitation of fast magnetoacoustic waves by phase mixing Alfvén waves in a cold plasma with a smooth inhomogeneity of density across a uniform magnetic field has been studied. If the fast waves were absent from the system, then nonlinearity leads to their excitation by transverse gradients in the Alfvén waves. The efficiency of the nonlinear Alfvén-fast magnetosonic wave coupling is strongly increased by the inhomogeneity of the medium. This nonlinear process suggests a mechanism of indirect plasma heating by phase mixing through the excitation of obliquely propagating fast waves. An important study on the nonlinear decay of phase-mixed Alfvén waves in the corona in the framework of two-fluid MHD has been attempted. It focuses on the parametric decay of the phase-mixed pump Alfvén mode into two Alfvén waves. This parametric decay is a nonlinear phenomenon that does not normally occur in ideal MHD. The parametric decay occurs for a relatively small amplitude and is more efficient than collisional damping.

In what follows, we shall briefly touch upon the concept of resonant absorption for both Alfvén and magnetoacoustic waves. As already mentioned in the phase-mixing scenario, we shall assume that the foot points of a set of field lines oscillate at a given frequency. The Alfvén frequency on each field line is different when we have a structured medium $(\omega(x))$. When the frequency of oscillation matches the local frequency of some continuum mode (ω_{excite}), the field line resonates and a large amplitude develops (see Fig. 4.8 for resonant absorption of Alfvén waves). Nonideal MHD limits the growth of the resonant mode and dissipates the incoming wave energy into heat along the entire length of the field line.

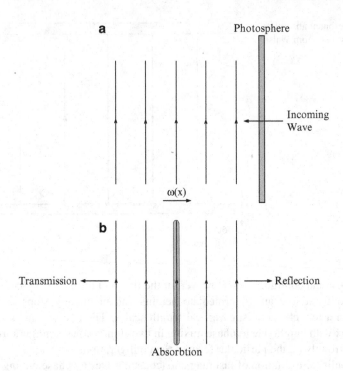

Fig. 4.9 Resonant absorption of magnetoacoustic waves; from Walsh (1999)

Regarding resonant absorption for magnetoacoustic waves, consider a structured medium such that each field line has its own Alfvén speed, with the exception that we shall investigate a wave incident on each side of the magnetic field (see Fig. 4.9 for the resonant absorption of magnetoacoustic waves). It is important that the pressure is continuous across the magnetic region. Thus, as the field line on the right side oscillates in response to the incoming wave, this oscillation will be transmitted to the next field line to the left, and the process will continue. As this coupling occurs across the field, it is possible that the field line oscillation will once again match some resonant mode, producing a resonant layer. This leads to transmission, absorption, and reflection of the incident oscillation.

The effect of velocity shear on the spectrum of MHD surface waves has been studied. A nonuniform intermediate region is taken into account, so that the magnetoacoustic surface wave can be subject to resonant absorption (Tirry et al. 1998). The dissipative solution has been derived analytically around the resonant surface in resistive MHD. Using these analytical solutions, the effect of velocity shear on the damping rate of the surface waves has been examined by substituting them in the eigenvalue code. The presence of the flow can both increase and decrease the efficiency of the resonant absorption. The resonance also leads to instability of the global surface mode for a certain range of values of the velocity shear. The resonant flow instabilities, which are different and distinct from the nonresonant

Kelvin–Helmholtz instabilities, can occur for velocity shears, which are well below the K–H threshold. The amplitude of the surface wave grows with time, in spite of the resonant absorption, which is a dissipative mechanism. The resonant flow instability has been explained in terms of negative energy waves, of which we will have something to discuss in the next chapter.

Nonlinear resonant interactions of different kinds of fast magnetosonic waves trapped in the inhomogeneity of a low-β plasma density, stretched along a magnetic field, has been studied (Erdelyi and Ballai 1999). A set of equations describing the amplitudes of interactive modes is derived for an arbitrary density profile. The decay instability of the wave with its highest frequency is in principle possible for such a system. If the amplitudes of interactive modes have close values, the long-period temporal and spatial oscillations will be part of the system. Dispersion relations of the fast magnetosonic waves trapped in the low$-\beta$ plasma slab with a parabolic transverse density profile have been evaluated. The interaction of kink and sausage waves has been studied. The spatial scale of a standing wave structure and the time spectrum radiation are formed due to the nonlinear interactions of loop modes, which contain information about the parameters of the plasma slab.

Linear dissipative MHD shows that driven MHD waves in a magnetic plasma with a high Reynolds number (ratio of inertial forces to viscous forces) exhibit a near-resonant behavior if the frequency of the wave becomes equal to the local Alfvén or slow frequency of a magnetic surface (Ruderman et al. 1997). This near-resonant behavior is confined to a thin dissipative layer, which embraces the resonant magnetic surface. Although the driven MHD waves have small amplitudes far away from the resonant magnetic surface, this near-resonant behavior in the dissipative layer may cause a breakdown of linear theory. The nonlinear behavior of driven MHD waves in the slow wave dissipative layer by the method of matched asymptotic expansions is utilized to determine the nonlinear evolution equation for the wave variables inside the dissipative layer. The absorption of the slow resonant wave in the dissipative layer generates a shear flow parallel to the magnetic surfaces with a characteristic velocity that depends on the amplitude of perturbations far away from the dissipative layer.

The effect of mass flow on resonant absorption and the overreflection of magnetoacoustic waves in low-β plasma with frequencies in the slow and the Alfvén continua have been studied Csik et al. (1998). They showed that in addition to the classical resonant absorption present in the plasma, driven resonant waves may also undergo overreflection. They also showed that relatively slow flows can have a very significant effect on the behavior of MHD waves.

Chapter 5
Waves in Nonuniform Media

5.1 Waves at a Magnetic Interface

In the previous chapter, we discussed the nature and properties of waves in a uniform medium. In this chapter, we shall study waves in a nonuniform medium, wherein we allow structures with finite geometries, such as the slab and the cylinder, and the magnetic field to have discontinuities in density and pressure. We will also study the behavior of waves in twisted magnetic fields.

Magnetic fields in general introduce structuring in the atmosphere of the Sun. They manifest as magnetic clumps, or flux tubes, at the photospheric level that are isolated from their neighboring environment (Spruit 1981; Spruit and Roberts 1983). Within the flux tubes, the field strengths are rather high, ranging from 1.5 kG (kilo Gauss) to 3 kG in sunspots (Cram and Thomas 1981). These flux tubes are the main building block of photospheric magnetic fields. They are found in the downdraughts of super granules and have a radius of about 100 km. Above the photospheric layers, in the low chromosphere, some of the isolated flux tubes rapidly expand to merge with their neighbors, sometimes completely filling the chromosphere and the corona with magnetic field.

Magnetic structuring is also present in the corona. The structuring and stratification of these magnetic field lines are more significant in the photosphere than in the corona because many of the structures (such as loops) have spatial scales that are less than or at most comparable with the pressure scale height.

To begin with, we examine waves in a magnetically structured atmosphere. The governing equations have been described in detail in the previous chapter. We shall briefly discuss the equilibrium state and the linearized equations of motion for completeness.

The equilibrium state for studying the oscillations is as follows:

$$\mathbf{B} = \mathbf{B}_0(x)\hat{z}, \quad \mathbf{p} = \mathbf{p}_0(x), \quad \rho = \rho_0(x), \quad T = T_0(x), \qquad (5.1)$$

A. Satya Narayanan, *An Introduction to Waves and Oscillations in the Sun*, Astronomy and Astrophysics Library, DOI 10.1007/978-1-4614-4400-8_5,
© Springer Science+Business Media New York 2013

where the pressure p_0, density ρ_0, and temperature T_0 are embedded in a unidirectional magnetic field B_0. All the equilibrium quantities are assumed to be functions of x in the Cartesian system, while the magnetic field is aligned along the z-axis. From the momentum equation, we have

$$\frac{d}{dx}\left[p_0 + \frac{B_0^2}{2\mu}\right] = 0, \tag{5.2}$$

implying that the total pressure (gas + magnetic) is constant.

Linear perturbations about the equilibrium on the equations of mass conservation, momentum, magnetic induction, and the equation of state (Roberts 1981, 1984) lead to

$$\frac{\partial \rho}{\partial t} + \nabla \cdot (\rho_0 \mathbf{v}) = 0 \tag{5.3}$$

$$\rho_0 \frac{\partial \mathbf{v}}{\partial t} = -\nabla\left[p + \frac{1}{\mu}\mathbf{B}_0 \cdot \mathbf{B}\right] + \frac{1}{\mu}(\mathbf{B}_0 \cdot \nabla)\mathbf{B} + \frac{1}{\mu}(\mathbf{B} \cdot \nabla)\mathbf{B}_0 \tag{5.4}$$

$$\frac{\partial \mathbf{B}}{\partial t} = \nabla \times (\mathbf{v} \times \mathbf{B}_0) \tag{5.5}$$

$$\nabla \cdot \mathbf{B} = 0 \tag{5.6}$$

$$\frac{\partial p}{\partial t} + \mathbf{v} \cdot \nabla p_0 = c^2 s\left[\frac{\partial \rho}{\partial t} + \mathbf{v} \cdot \nabla \rho_0\right], \tag{5.7}$$

describing disturbances that have a small amplitude (linear) with velocity $\mathbf{v} = (v_x, v_y, v_z)$ and magnetic field $\mathbf{B} = (B_x, B_y, B_z)$, pressure p, and density ρ.

The magnetic pressure is given by $p_m = \frac{1}{\mu}(\mathbf{B}_0 \cdot \mathbf{B})$ and the total pressure by $p_T = p + p_m$. The linearized equations of motion reduce to

$$\rho_0\left[\frac{\partial^2}{\partial t^2} - V_A^2 \frac{\partial}{\partial z^2}\right]\mathbf{v}_\perp + \mathbf{v}_\perp\left[\frac{\partial p_T}{\partial t}\right] = 0 \tag{5.8}$$

and

$$\left[\frac{\partial^2}{\partial t^2} - c_s^2 \frac{\partial^2}{\partial z^2}\right]\mathbf{v}_\parallel - c_s^2 \nabla \cdot \left[\frac{\partial \mathbf{v}_\perp}{\partial z}\right] = 0. \tag{5.9}$$

Here, $\mathbf{v}_\perp = (v_x, v_y, 0)$ and $v_\parallel = (0, 0, v_z)$ are the velocities perpendicular and parallel to the applied magnetic field $\mathbf{B}_0$. The velocity components are related to the pressure perturbations p_T and p_m as

$$\frac{\partial p_T}{\partial t} = -\rho_0\left[c_s^2 \frac{\partial v_\parallel}{\partial z} + (c_s^2 + V_A^2)\nabla \cdot \mathbf{v}_\perp\right] \tag{5.10}$$

$$\frac{\partial p_m}{\partial t} = \left[\frac{dp_0}{dx}\right]v_x - \rho_0 V_A^2 \nabla \cdot \mathbf{v}_\perp. \tag{5.11}$$

In the above relations, $c_s(x) = (\gamma p_0(x)/\rho_0(x))^{1/2}$ and $V_A(x) = (\mathbf{B}_0(x)/(\mu\rho_0(x)))^{1/2}$ are the sound speed and Alfvén speed, respectively.

Introducing sinusoidal disturbance as

$$v_x = v_x(x)e^{i(\omega t - k_y y - k_z z)},$$

one can eliminate the velocity components v_y and v_z, to have equations only in terms of $v_x(x)$ and $p_T(x)$, to derive a pair of ordinary differential equations as follows:

$$\frac{dp_T}{dx} + \frac{i\rho_0}{\omega}(\omega^2 - k_z^2 V_A^2)v_x = 0 \qquad (5.12)$$

$$(\omega^2 - k_z^2 V_A^2)(\omega^2 - k_z^2)\frac{dv_x}{dx} + \frac{i\omega}{\rho_0(c_s^2 + V_A^2)}(\omega^2 - \omega_S^2(x))(\omega^2 - \omega_f^2(x))p_T = 0,$$
$$(5.13)$$

where

$$\omega_S^2 + \omega_f^2 = (k_y^2 + k_z^2)(c_s^2 + V_A^2)$$

and

$$\omega_S^2 \omega_f^2 = k_z^2(k_y^2 + k_z^2)c_s^2 V_A^2.$$

ω_S and ω_f are the frequencies corresponding to the slow and fast modes, respectively. At this stage, we define the cusp speed c_T as

$$c_T(x) = c_s V_A / (c_s^2 + V_A^2)^{1/2},$$

which is subsonic (Mach number less than 1) and sub-Alfvénic (Alfvénic Mach number less than 1). The pair of ordinary differential equations can be reduced by eliminating either p_T or v_x. Eliminating p_T gives

$$\frac{d}{dx}\left[\frac{\rho_0(c_s^2 + V_A^2)(\omega^2 - k_z^2 V_A^2)(\omega^2 - k_z^2 c_T^2)}{(\omega^2 - \omega_S^2)(\omega^2 - \omega_f^2)}\frac{dv_x}{dx}\right] + \rho_0(\omega^2 - k_z^2 V_A^2)v_x = 0. \quad (5.14)$$

Equation (5.14) may be considered the wave equation for MHD waves in an inhomogeneous medium. Looking at the equation, it is clear to us that it has two singularities: an Alfvén singularity $\omega^2 = k_z^2 V_A^2(x)$ and the cusp singularity $\omega^2 = k_z^2 c_T^2(x)$. These singularities may be associated with the occurrence of continuous spectra and may be attributed to the highly anisotropic nature of the Alfvén and magnetosonic waves. There are two other expressions that are cutoff points and not singularities, given by $\omega^2 = \omega_S^2(x)$ and $\omega^2 = \omega_f^2(x)$. Corresponding to wave reflection or trapping, these are associated with a change from oscillatory to evanescent wave behavior.

One can think of two types of discretely structured media, one possessing a single interface separating two regions with different plasma parameters, such as density, pressure, and magnetic field, and the other in which there are two interfaces combining to form a slab or tube of magnetic field that in some sense is different from its environment.

We shall start our discussion of the waves at a magnetic interface with the simple case of an incompressible fluid (the sound velocity approaching ∞). Subsequently, we shall generalize this to include the effects of compressibility, flows, viscosity, nonparallel propagation, and gravity.

One of the earliest works on hydromagnetic surface waves was by Wentzel (1979). It dealt with the nature of surface waves, with emphasis on the dispersions, the spatial extent, and the degree of compression and coupling with hydromagnetic waves. It is interesting to note that surface waves involve finite gas compression. Subsequently, Rae and Roberts (1983) studied MHD wave motion in a magnetically structured atmosphere. For an equilibrium at resonance, the wave energy possesses infinite solutions. However, for certain conditions, the wave energy is transmitted through magnetoacoustic boundaries.

Incompressible medium: In the limit of $c_s \to \infty$ and setting k_y and v_y equal to 0 for simplicity, for two-dimensional motions $(v_x, 0, v_z)$, the wave equation (5.14) reduces to

$$\frac{d}{dx}\left[\rho_0(k_z^2 V_A^2 - \omega^2)\frac{dv_x}{dx}\right] - k_z^2 \rho_0(k_z^2 V_A^2 - \omega^2)v_x = 0. \qquad (5.15)$$

We shall divide the magnetic interface being made up of two media $(x < 0)$ and $(x > 0)$ with the following assumptions:

$$B_0(x) = B_e \quad for \quad x > 0$$

$$B_0(x) = B_0 \quad for \quad x < 0.$$

A similar assumption will be made for the pressure and density. All the values for the pressure, density, and magnetic field are assumed to be uniform with a discontinuity at the interface $x = 0$.

The total pressure balance is given by the relation

$$p_0 + \frac{B_0^2}{2\mu} = p_e + \frac{B_e^2}{2\mu}, \qquad (5.16)$$

where p_e and B_e are the pressure and magnetic field for the medium $x > 0$. Combining the ideal gas law, the above relation can be simplified to yield

$$\rho_e(c_e^2 + (1/2)\gamma V_{Ae}^2) = \rho_0(c_0^2 + (1/2)\gamma V_A^2). \qquad (5.17)$$

c_0, c_e, V_A, and V_{Ae} are, respectively, the sound and Alfvén speeds in the media $x < 0$ and $x > 0$. With the above assumptions, the wave equation reduces to

$$\rho_0(k_z^2 V_A^2 - \omega^2)\left[\frac{d^2 v_x}{dx^2} - k_z^2 v_x\right] = 0. \qquad (5.18)$$

The above equation is valid in both media. We are not interested in the vanishing of $\omega^2 = k_z^2 V_A^2$, but only in the differential operator

$$\frac{d^2 v_x}{dx^2} - k_z^2 v_x = 0. \tag{5.19}$$

The above equation is easy to solve. It possesses a very simple exponential type of solutions, such as

$$v_x \sim e^{-k_z x}$$

and

$$v_x \sim e^{k_z x}.$$

The wave equation being linear, a linear combination of the above solutions will also be a solution. However, we impose the condition that v_x is bounded for $x \to \pm\infty$. We take the solution as

$$v_x(x) = \alpha_e e^{-k_z x} \quad x > 0$$

$$v_x(x) = \alpha_0 e^{k_z x} \quad x < 0, \tag{5.20}$$

satisfying the boundedness of v_x for $|x| \to \infty$. α_0 and α_e are arbitrary constants.

In order to derive the dispersion relation, we have to impose the boundary conditions at the interface $x = 0$. The two conditions are the continuity of the normal velocity component v_x and the pressure perturbation at the interface. The pressure continuity condition leads to

$$p_T \sim \rho_0(x)(k_z^2 V_A^2(x) - \omega^2)\frac{dv_x}{dx} \tag{5.21}$$

being continuous at $x = 0$. Combining both conditions, the dispersion relation reduces to

$$\omega^2 = k_z^2 c_k^2 \equiv k_z^2 \left[\frac{\rho_0 V_A^2 + \rho_e V_{Ae}^2}{\rho_0 + \rho_e} \right]. \tag{5.22}$$

The dispersion relation (5.22) describes hydromagnetic surface waves. It is evident from the relation that the frequency of these waves depends on the magnitude of the density and the Alfvén velocities in both media, implying that a change in the magnetic field or density will result in the propagation of these waves. The wave is confined to the interface with a penetration depth given by k_z^{-1}.

Wave propagation in a magnetically structured configuration has been studied by several authors (Roberts 1981a; Somasundaram and Uberoi 1982; Miles and Roberts 1989; Jain and Roberts 1991; Uberoi and Satya Narayanan 1986; Singh and Talwar 1993; Uberoi and Satya Narayanan 1986), who investigated the properties of wave arising on a single magnetic interface.

Compressible medium: In this case, as mentioned in the previous chapter, one would expect the appearance of two additional modes, the magnetosonic modes. Returning to Eq. (5.14) with the restriction of two-dimensionality $(v_x, 0.v_z)$, we have

$$\frac{d}{dx}\left[\frac{\rho_0(x)(c_s^2(x)+V_A^2(x))(k_z^2 c_T^2(x)-\omega^2)}{(k_z^2 c_s^2(x)-\omega^2)}\frac{dv_x}{dx}\right] - \rho_0(x)(k_z^2 V_A^2(x)-\omega^2)v_x = 0. $$

$$(5.23)$$

The relation between p_T and v_x is given by

$$p_T = \frac{i\rho_0(x)}{\omega}(c_s^2(x)+V_A^2(x))\frac{(k_z^2 c_T^2(x)-\omega^2)}{(k_z^2 c_s^2(x)-\omega^2)}\frac{dv_x}{dx}. \tag{5.24}$$

In a uniform medium, Eq. (5.23) reduces to

$$\frac{d^2 v_x}{dx^2} - m_0^2 v_x = 0, \tag{5.25}$$

where

$$m_0^2 = \frac{(k_z^2 c_s^2 - \omega^2)(k_z^2 V_A^2 - \omega^2)}{(c_s^2 + V_A^2)(k_z^2 c_T^2 - \omega^2)}. \tag{5.26}$$

The above expression reduces to k_z^2 for the incompressible case.

Let's first consider the single interface model as earlier. Continuing the analysis, similar to the incompressible case wherein we insist that the mode remain bounded for large values of x and we apply the boundary conditions across $x = 0$, yields the dispersion relation

$$\frac{\omega^2}{k_z^2} = V_A^2 - \left[\frac{M}{1+M}\right](V_A^2 - V_{Ae}^2), \tag{5.27}$$

where $M = \rho_e m_0/\rho_0 m_e$. Also, m_e is the value of m_0 in the region $x > 0$. The above dispersion relation describes the behavior of magnetoacoustic surface waves, propagating along the interface $x = 0$. It is important to realize that the dispersion relation mentioned above is transcendental as M is a function of ω, unlike the case of incompressible medium, where the phase speed could be written explicitly in terms of the wavenumbers and the interfacial parameters. Thus, it may possess more than one solution, for example, the fast and slow magnetoacoustic surface waves, due to compressibility.

If one side of the interface is field-free, say $B_e = 0$, then the above equation reduces to

$$(k_z^2 V_A^2 - \omega^2)m_e = \frac{\rho_e}{\rho_0}\omega^2 m_0, \tag{5.28}$$

where now

$$m_e^2 = k_z^2 - \frac{\omega^2}{c_e^2}. $$

The dispersion relation for hydromagnetic surface waves along the interface between two compressible plasma media for general values of the different parameters arising in the model when the magnetic fields across the interface vary both in direction and in magnitude can be simplified to yield (Uberoi and Satya Narayanan 1986)

$$\varepsilon_1(k,\omega)(m_e^2 + l^2)^{1/2} + \varepsilon_2(k,\omega)(m_0^2 + l^2) = 0. \tag{5.29}$$

Here

$$\varepsilon_{1,2}(k,\omega) = k^2 B_{0,e}^2/\mu - \rho_{0,e}\omega^2.$$

The propagation characteristics of both the slow and fast modes show variations with the angle between the magnetic field directions on either side of the interface. There exists a critical angle between the magnetic fields for which the propagation band for surface waves becomes zero and both modes propagate with bulk Alfvén velocity on either side of the media, with negligible compressibility effects. When finding roots of the transcendental equation, which represent the possible modes of surface wave propagation, it should be noted that the dispersion relation will have real roots only when ε_1 and ε_2 are of opposite sign, and for an exponentially decaying solutions, the terms $M_{0,e}^2 + \tan^2\theta$ should both be positive, where $M_{0,e} = m_{0,e}/k$, and $\tan\theta = l/k$.

Considering the phase speed and sound speed plane, ε_1 and ε_2 are of opposite signs only when

$$\min(V_{A1,2}) < \omega/k < \max(V_{A1,2}).$$

The possible regions of surface wave propagation are shown in Fig. 5.1.

5.2 Surface and Interfacial Waves

In the previous section, we studied the properties of the surface waves when both the wavenumber vector and the magnetic fields on either side of the interface were parallel to the surface $x = 0$. In this section, we shall extend the above results to include the case when the wavenumber and magnetic field vary both in magnitude and in direction. A simple two-layer model to investigate the surface waves arising due to the interaction between two fluids of different densities has been studied Satya Narayanan (1996a). The upper fluid is under the influence of a magnetic field inclined at an angle to the interface ,while the lower fluid is field-free. The wave vector is in a direction different from the magnetic fluid.

The dispersive characteristics of interfacial waves in low-β plasma have been studied Satya Narayanan (1996a). The condition for the existence of these waves was derived. The wavenumber and magnetic field have the following form:

$$\mathbf{k} = (0, k\sin\theta, k\cos\theta)$$

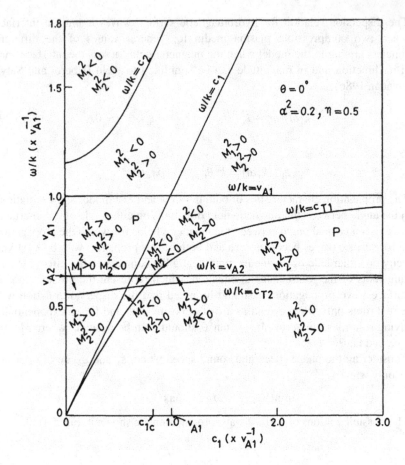

Fig. 5.1 Possible regions for surface wave propagation for specific values of the interface parameters; from Somasundaram and Uberoi (1982)

and

$$\mathbf{B}_{01,2} = (0, B_{01,2}\cos\gamma_{1,2}, B_{01,2}\sin\gamma_{1,2}).$$

Taking ρ_0 and ρ_e to be the densities, $B_{01,2}$, the magnetic fields (constant) on either side of the interface, and c_s and V_A as the sound and Alfvén speeds, the dispersion relation by substituting the boundary conditions at the interface can be written as

$$\tau_1 \varepsilon_1(\omega,k) + \tau_2 \varepsilon_2(\omega,k) = 0, \tag{5.30}$$

where

$$\varepsilon_1(\omega,k) = \rho_e(-\omega^2 + k^2 V_{Ae}^2 \sin^2(\theta + \gamma_2)) \tag{5.31}$$

$$\varepsilon_2(\omega,k) = \rho_0(-\omega^2 + k^2 V_A^2 \sin^2(\theta + \gamma_1)), \tag{5.32}$$

and $\tau_{1,2}$ are given by

$$\tau_{1,2}^2 = \frac{\omega^4 - k^2\sin^2(\theta + \gamma_{1,2})(V_{Ae,A}^2 + c_{se,s0}^2) + k^4 c_{se,s0}^2 V_{Ae,A}^2 \sin^4(\theta + \gamma_{1,2})}{k^2 c_{se,s0}^2 V_{Ae,A}^2 \sin^2(\theta + \gamma_{1,2}) - \omega^2(c_{se,s0}^2 + V_{Ae,A}^2)}. \quad (5.33)$$

Let's consider the incompressible limit of the above relation. We'll set $c_{se,s0} \to \infty$, $\tau_{1,2}^2 \to k^2$, so that the dispersion relation reduces to

$$\rho_0(-\omega^2 + k^2 V_A^2 \sin^2(\theta + \gamma_1)) + \rho_e(-\omega^2 + k^2 V_{Ae}^2 \sin^2(\theta + \gamma_2)) = 0, \quad (5.34)$$

which gives the analytical expression for the phase speed as

$$\frac{\omega^2}{k^2} = \frac{B_{01}^2 \sin^2(\theta + \gamma_1) + B_{02}^2 \sin^2(\theta + \gamma_2)}{(\rho_0 + \rho_e)}. \quad (5.35)$$

We'll introduce the nondimensional parameters

$$\eta = \rho_0/\rho_e, \quad \alpha = B_{02}/B_{01}, \quad \omega/kV_A = y, \quad c_{s0}/V_A = x$$

and write $(\theta + \gamma_1) = \phi$, and $(\gamma_1 - \gamma_2) = \chi$.

As already mentioned, the dispersion relation will have real roots only when ε_1 and ε_2 have opposite signs, and $\tau_{1,2}$ should both be positive for the roots to represent surface wave propagation. This implies that for positive roots, ω/k should lie in the range

$$\min(V_{Ae,A}\sin(\theta + \gamma_{1,2})) < \omega/k < \max(V_{Ae,A}\sin(\theta + \gamma_{1,2})).$$

By simple algebra, one can show that

$$\varepsilon_1 = (-y^2 + \alpha^2\eta\sin^2(\phi - \chi)) = (-y^2 + y_1^2)$$

$$\varepsilon_2 = \eta(-y^2 + \sin^2\phi) = (-y^2 + y_2^2)\eta$$

$$\tau_1^2 = \frac{y^4 - y^2(1 + x^2) + x^2\sin^2\phi}{x^2\sin^2\phi - y^2(1 + x^2)} = \frac{(y^2 - y_4^2)(y^2 - y_5^2)}{(y_3^2 - y^2)}$$

$$\tau_2^2 = \frac{y^4 - y^2(\alpha^2\eta + x^2(c_{s2}^2/c_{s1}^2)) + (c_{s2}^2/c_{s1}^2)x^2\alpha^2\sin^2(\phi - \chi)}{(c_{s2}^2/c_{s1}^2)\alpha^2\eta x^2\sin^2(\phi - \chi) - y^2(\alpha^2\eta + x^2(c_{s2}^2/c_{s1}^2))}$$

$$= \frac{(y^2 - y_7^2)(y^2 - y_8^2)}{(y_6^2 - y^2)}.$$

Here the y_i are functions of the angles, the sound, and Alfvén velocities in both media, respectively.

Let's consider the special case of $c_{s1}/V_A \ll 1$, $B_{01} = 0$, and $\gamma_1 = \gamma_2$. The expressions in the dispersion relation (5.30) reduce to

$$\varepsilon_1 = \rho_0(-\omega^2 + k^2 V_A^2 \sin^2(\theta + \gamma))$$

$$\varepsilon_2 = \rho_e(-\omega^2)$$

$$\tau_1^2 = k^2 \left(1 - \frac{\omega^2}{k^2 V_A^2}\right)$$

$$\tau_2^2 = k^2 \left(1 - \frac{\omega^2}{k^2 c_{se}^2}\right).$$

Introducing the nondimensional quantities $\alpha = \rho_e/\rho_0$, $\delta = c_{s2}/V_A$, $x = \omega/kV_A$, and simplifying the relation yield

$$y^3 + Ay^2 + By + C = 0. \tag{5.36}$$

A, B, and C are given by

$$A = \frac{\delta^2(\alpha^2 - 1) - 2\sin^2(\delta + \gamma)}{(1 - \alpha^2\delta^2)}$$

$$B = \frac{(1 + 2\delta^2)\sin^2(\theta + \gamma)}{(1 - \alpha^2\delta^2)}$$

$$C = \frac{-\delta^2\sin^4(\theta + \gamma)}{(1 - \alpha^2\delta^2)}.$$

The condition for the existence of surface waves reduces to

$$\text{Min}(V_A\sin(\theta + \gamma)) < \omega/k < \text{Max}(V_A\sin(\theta + \gamma)). \tag{5.37}$$

In what follows, let's assume that $\gamma_1 \neq \gamma_2$, with the same expression for wavenumber and magnetic field as mentioned at the beginning of the chapter.

The dispersion relation (5.30) will have coefficients as follows:

$$\varepsilon_1(\omega, k) = \rho_0(-\omega^2 + k^2 V_A^2 \sin^2(\theta + \gamma_1)) \tag{5.38}$$

$$\varepsilon_2(\omega, k) = \rho_e(-\omega^2 + k^2 V_{Ae}^2 \sin^2(\theta + \gamma_2)) \tag{5.39}$$

and

$$\tau_{1,2}^2 = (\omega^4 - A + B)/(C - D), \tag{5.40}$$

where

$$A = k^2\omega^2(V_{A,Ae}^2 + c_{s0,se}^2)$$

$$B = k^4 c_{s0,se}^2 V_{A,Ae}^2 \sin^2(\theta + \gamma_{1,2})$$

$$C = k^2 c_{s0,se}^2 V_{A,Ae}^2 \sin^2(\theta + \gamma_{1,2})$$

$$D = \omega^2 \left(c_{s0,se}^2 + V_{A,Ae}^2\right).$$

Some special cases follow:
(Low-β plasma): The expressions for $\tau_{1,2}$ reduce to

$$\tau_{1,2}^2 = k^2 \left(1 - \frac{\omega^2}{k^2 V_{A,Ae}^2}\right).$$

For the low-β case, the pressure balance condition at the interface will yield

$$\rho_0 V_A^2 \approx \rho_e V_{Ae}^2.$$

Introducing the nondimensional parameters as mentioned earlier, the relation reduces to

$$(1 - \alpha x^2)^{1/2}(\lambda_1^2 - x^2) + (1 - x^2)^{1/2}(\lambda_2^2 - \alpha x^2) = 0, \qquad (5.41)$$

where $\lambda_{1,2} = \sin(\theta + \gamma_{1,2})$.
The case when $\gamma_1 = \gamma_2 = \pi/2$: The relation simplifies to

$$(1 - \alpha x^2)^{1/2}(\cos^2\theta - x^2) + (1 - x^2)^{1/2}(\cos^2\theta - \alpha x) = 0, \qquad (5.42)$$

which can be further simplified to yield

$$\alpha x^4 - (1 + \alpha)x^2 + \cos^2\theta(1 + \sin^2\theta) = 0,$$

so that

$$x^2 = ((1 + \alpha) \pm [(1 - \alpha)^2 + 4\alpha\sin^4\theta]^{1/2}/2\alpha.$$

The case when $\theta = 0$: This refers to a parallel propagation, and the dispersion relation can be simplified to yield

$$x^6(\alpha^2 - \alpha) + x^4(1 - \alpha^2 + 2\alpha\sin^2\gamma_1 - 2\alpha\sin^2\gamma_2)x^2(\sin^4\gamma_2 - \alpha\sin^4\gamma_1$$
$$+ 2\alpha\sin^2\gamma_2 - 2\sin^2\gamma_1) + (\sin^4\gamma_1 - \sin^4\gamma_2) = 0. \qquad (5.43)$$

The case when $\gamma_1 = \gamma_2 \neq \pi/2$: The solution of the dispersion relation in this case can be written as

$$x^2 = ((1 + \alpha) \pm [(1 - \alpha)^2 + 4\alpha\cos^4\gamma]^{1/2})/2\alpha.$$

Finally, the interesting case when the sum of the angles $\theta + \gamma_1$ and $\theta + \gamma_2$ is $\pi/2$, the relation reduces to

$$(1 - \alpha x^2)^{1/2}(1 - x^2) + (1 - x^2)^{1/2}(1 - \alpha x^2) = 0. \qquad (5.44)$$

Presence of steady flows: The combined effect of nonparallel propagation and steady flow on the properties of hydromagnetic waves (Joarder and Satya Narayanan

2000) will be discussed below. It can in principle give rise to backward-propagating surface waves that may be subject to negative energy instabilities. The basic magnetic field is (B_e, B_0), while the steady flow has the form (U_e, U_0) with constant values for magnetic field and velocity shear. The dispersion relation is similar to the one discussed in Eq. (5.30), except that the frequency is altered due to the flow. The dispersion relation looks like

$$\rho_e(V_{Ae}^2 - \Omega_e^2)m_0 + \rho_0(V_A^2 - \Omega_0^2)m_e = 0. \tag{5.45}$$

Here

$$\Omega_e = \omega - kU_e, \qquad \Omega_0 = \omega - kU_0.$$

The condition for the existence of surface waves in the presence of flow is given by

$$\text{Max}(v_{ce}^-(\theta), V_A - |U_0|) < c < \text{Min}(v_{c0}^+(\theta) - |U_0|, V_{Ae}),$$

where

$$v_{c(e,0)}^\pm(\theta) = \left[(1/2)(V_{A,Ae}^2 + c_{s0,se}^2)\sec^2\theta \pm (1/2)\left[(V_{A,Ae}^2 + c_{s0,se}^2)^2\sec^4\theta \right.\right.$$

$$\left.\left. -4V_{A,Ae}^2 c_{s0,se}^2 \sec^2\theta \right]^{1/2} \right]^{1/2}. \tag{5.46}$$

Here the "+" refers to fast waves, while the "−" refers to that of slow waves, and $c = \omega/k$.

In the limit of $\theta \to \pi/2$, the phase speed of the surface wave (fast) takes the form

$$c = -\frac{\rho_0}{(\rho_0 + \rho_e)}|U_0| + \left[\frac{\rho_0 V^2 A + \rho_e V_{Ae}^2}{(\rho_0 + \rho_e)} - \frac{\rho_0 \rho_e}{(\rho_0 + \rho_e)^2}|U_0^2| \right]^{1/2}. \tag{5.47}$$

The phase speed of the forward-propagating slow surface wave has the form (in the limit $\theta \to \pi/2$) given by

$$c_a \approx -\left[\frac{\rho_e}{(\rho_0 + \rho_e)} \right]|U_e| + \left[\frac{\rho_0}{(\rho_0 + \rho_e)}V_A^2 - \frac{\rho_0 \rho_e}{(\rho_0 + \rho_e)^2}|U_e|^2 \right]^{1/2}. \tag{5.48}$$

Tangential discontinuity with inclined fields and flows: The combined effect of nonparallel propagation, steady flow, and inclined magnetic fields on either side of a polar tangential discontinuity will be examined, with a change in the field strength of the magnetic field, though uniform in each layer. The density is also assumed to be different on both sides of the interfacial layer. The interface will in principle support both body waves as well as surface waves (Satya Narayanan and Ramesh 2002; Joarder and Nakariakov 2006; Satya Narayanan et al. 2008; Joarder et al. 2009). This will also support fast, Alfvén, and slow modes depending on the parametric values of the system.

The equilibrium is such that

$$\frac{d}{dx}\left(p_0 + \frac{B_0^2}{2\mu}\right) = 0.$$

The perturbations are as follows:

$$\bar{\rho} = \rho_0(x) + \rho; \quad \bar{\mathbf{v}} = \mathbf{U}(x) + \mathbf{v}; \quad \bar{p} = p_0(x) + P; \quad \bar{\mathbf{B}} = \mathbf{B}(x) + \mathbf{b},$$

where $\mathbf{U} = (0, U_y, U_z)$, $\mathbf{B} = (0, B_y, B_z)$. The basic equations of MHD can be simplified to get a single differential equation for the velocity component v_x as

$$v_x'' + (m_0^2 + k_y^2)v_x = 0, \tag{5.49}$$

with

$$m_0^2 = \frac{\Omega^2 + k_x^2 c_s^2\left(\omega_T^2 - \Omega^2\right)}{c_s^2\left(\omega_T^2 - \Omega^2\right)},$$

where Ω, ω, k_x, k_y, and c_s are the Doppler-shifted frequency, angular frequency, wavenumbers, and sound speed, respectively. We shall assume that the variation of the wavenumber is different from that of the magnetic field; that is,

$$\mathbf{k} = (0, k\sin\theta, k\cos\theta) \quad \mathbf{B} = (0, B\sin\gamma, B\cos\gamma).$$

Using similar arguments, the dispersion relation can be shown to be

$$\rho_0\left[k^2 c_{s0}^2 \cos^2(\theta - \gamma) - \Omega_0^2\right](m_e^2 + k_y^2)^{1/2} + \rho_e\left[k^2 c_{se}^2 \cos^2(\theta - \gamma) - \Omega_e^2\right](m_0^2 + k_y^2)^{1/2} = 0. \tag{5.50}$$

Introducing the following nondimensional variables:

$$\alpha = \frac{\rho_e}{\rho_0}, \quad \delta = \frac{U_e}{U_0}, \quad \varepsilon = \frac{U_0}{V_A}, \quad x = \frac{\omega}{kV_A}$$

and for low-β plasma, the dispersion relation reduces to

$$\left[\cos^2(\theta - \gamma) - (x - \varepsilon)^2\right]\left[1 - \alpha(x - \varepsilon\delta)^2\right]^{1/2}$$
$$+ \left[\cos^2(\theta - \gamma) - \alpha(x - \varepsilon\delta)^2\right]\left[1 - (x - \varepsilon)^2\right]^{1/2}. \tag{5.51}$$

The dispersive characteristics of surface waves with flows was studied by Satya Narayanan et al. (2008), and the solution of the dispersion relation is presented in Figs. 5.2 and 5.3.

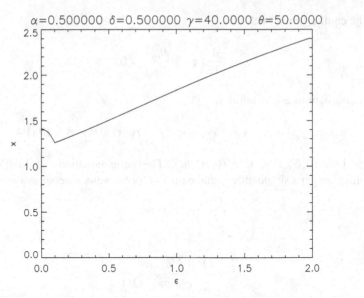

Fig. 5.2 Dispersive characteristics of surface waves with flows, for some specific parametric values; from Satya Narayanan et al. (2008)

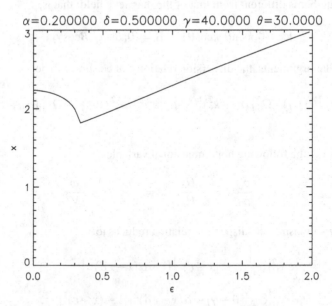

Fig. 5.3 Same as in the previous figure, for different parametric values; from Satya Narayanan et al. (2008)

Two-mode structure of Alfvén surface waves: The Alfvén surface waves propagating along a viscous conducting fluid–vacuum interface will be discussed (Uberoi and Somasundaram 1982). In addition to the ordinary Alfvén surface wave, modified by viscous effects, the interface can support a second mode, which is the damped solution of the dispersion equation. The equations of motion for an incompressible viscous fluid of mass density ρ_0, embedded in an external magnetic field $\mathbf{B}_0$, with small perturbations from the equilibrium state can be written as a coupled system of differential equations in v_x and v_z as follows:

$$kD(D^2 - \tau^2)v_x + iK^2(D^2 - \tau^2)v_z = 0 \tag{5.52}$$

$$k(D^2 - \tau^2)v_x + iD(D^2 - \tau^2)v_z = 0, \tag{5.53}$$

where $D = d/dx$, $\tau^2 = K^2 - (i\rho_0/v\omega)(\omega^2 - k^2 V_A^2)$, and $K^2 = k^2 + l^2$. The above equations can be simplified to yield

$$(D^2 - \tau^2)(D^2 - K^2)v_x = 0. \tag{5.54}$$

In addition to the boundary conditions we mentioned earlier, namely, the continuity of (1) the normal velocity component, (2) the total pressure continuous across the boundary, we must insist on two more boundary conditions, namely, (3) the continuity of the tangential velocity and (4) the tangential viscous stress. Without loss of generality, we assume that $l = 0$, the density and viscosity to be negligible at the upper portion of the interface, say $x > 0$. The dispersion relation can be simplified and written as

$$(x^2 - 1)^2 - \alpha^2(x^2 - 1) + i4v(x^2 - 1) + 4v^2 T_1 - 4v^2 = 0, \tag{5.55}$$

where

$$x = \frac{\omega}{kV_A}, \quad V = \frac{v\omega}{\rho_0 V_A^2}, \quad T_1 = k\left(1 - \frac{i(x^2 - 1)}{V}\right)^{1/2}.$$

Squaring (5.55) and taking the common factor $(x^2 - 1)$, which represents the bulk mode outside, we have

$$x^6 + x^4(-3 - 2\alpha^2 + i8V) + x^2(1 + \alpha^2)[(1 + \alpha^2) + 2(1 - 4iV)] - 8V(i + 3V)]$$
$$+ (1 + \alpha^2)[-1 - \alpha^2 + 8V(i + V)] + 16V(1 - iV) = 0. \tag{5.56}$$

The effect of uniform flows on the viscous damping of Alfvén surface waves at a tangential discontinuity will be a generalization of the above study with the flows included (David Rathinavelu et al. 2009, 2010). In this case, the angular frequency is modified to a Doppler-shifted frequency. Introducing the nondimensional parameters as follows:

$$\beta = \frac{B_e}{B_0}, \quad x = \frac{\omega}{kV_A}, \quad R = \frac{U}{V_A}, \quad v = \frac{vk}{\rho_0 V_A},$$

Fig. 5.4 Solution of the dispersion relation for damped Alfvén mode with $\beta^2 = 0.02$ and R = 0.0; from Rathinavelu, Sivaraman and Satya Narayanan (2010)

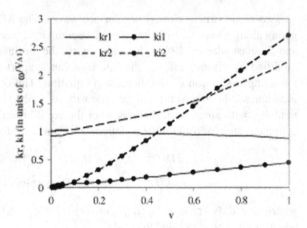

the magnetic field ratio, normalized phase velocity, flow velocity, and viscosity, respectively, the following relation for the damped Alfvén mode (in addition to the surface mode) is written as

$$(x-R)^6 + C(x-R)^5 + D(x-R)^4 + E(x-R)^3 + F(x-R)^2 + G(x-R) + H = 0,$$
$$(5.57)$$

where

$$C = 6iv$$
$$D = 2ivx - 9v^2 - 2\beta^2 - 3$$
$$E = -14v^2x - 12iv - 6iv\beta^2 - 4v^2R$$
$$F = -v^2x^2 - 2iv\beta^2x - 4ivx - 24iv^3x + 9v^2 + 2(1+\beta^2) + (1+\beta^2)^2$$
$$G = 8iv^3x^3 + 6v^2x + 8v^2x(1+\beta^2) + 12iv^3Rx + 4iv^3R^2 + (6iv + 4v^2R)(1+\beta^2)$$
$$H = v^2x^2 + 2ivx(1+\beta^2) - (1+\beta^2)^2.$$

The dispersion relation has been solved numerically. In the absence of flow, one clearly observes the two-mode structure of Alfvén surface waves (Fig. 5.4). However, when the flow is introduced $(R > 0)$, the second mode becomes evanescent after a critical value of v and a new mode appears at a higher value of v (see Fig. 5.5). It is observed that the flow suppresses the existing modes and supports the evolution of new modes.

Magnetoacoustic-gravity surface waves with flows: The linear theory of parallel propagation of magnetoacoustic-gravity (MAG) surface waves for an interface of a plasma embedded in a horizontal magnetic field above a field-free steady plasma medium will be discussed below (Erdelyi and Ballai 1999; Varga and Erdelyi 2001a,b). The dispersion relation is derived and studied for the case of constant Alfvén speed . The presence of new modes called flow or v-modes is

Fig. 5.5 Same as in the previous figure for $R = 1.5$; from Rathinavelu, Sivaraman, and Satya Narayanan (2010)

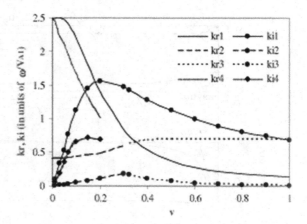

observed as a consequence of steady flows. The equilibrium of magnetohydrostatics is given by

$$\frac{d}{dz}\left(p(z) + \frac{B^2(z)}{2\mu}\right) = -g\rho(z). \tag{5.58}$$

In both regions, the assumption of constant Alfvén velocity implies

$$\rho(z) = \rho_0 \exp(-z/H_B), \quad z > 0$$
$$\rho(z) = \rho_e \exp(-z/H_e), \quad z < 0,$$

where H_B and H_e are the density scale heights in both regions, respectively. Two-dimensional linear, isentropic perturbations reduce the linearized compressible ideal MHD equations into a single ordinary differential equation (ODE) for velocity component v_z:

$$\frac{d}{dz}\left[\frac{\rho(V_A^2 + c_s^2)(\Omega^2 - k_x^2 c_T^2)}{k_x^2 c_s^2 - \Omega^2}\frac{dv_z}{dz}\right]$$

$$-\left[\rho\left(\Omega^2 - k_x^2 V_A^2\right) + \frac{g^2 k_x^2 \rho}{k_x^2 c_s^2 - \Omega^2} + g k_x^2 \frac{d}{dz}\left(\frac{\rho c_s^2}{k_x^2 c_s^2 - \Omega^2}\right)\right]v_z = 0. \tag{5.59}$$

Applying the usual boundary conditions, the dispersion relation for MAG can be written as

$$\frac{\omega^2}{k_x^2} = \frac{\rho_0 V_A^2}{\rho + \frac{\Omega^2}{\omega^2}\rho_e\frac{m_0^2(M_e + 1/2H_e)}{m_e^2(M_0 - 1/2H_B)}} - g\frac{\frac{\rho_0 c_s^2}{k_x^2 c_s^2 - \omega^2} - \frac{\rho_0 c_s^2}{k_x^2 c_s^2 - \Omega^2}}{\frac{\rho_0(M_0 - 1/2H_B)}{m_0^2} + \frac{\Omega^2}{\omega^2}\frac{\rho(M_e + 1/2H_e)}{m_e^2}}, \tag{5.60}$$

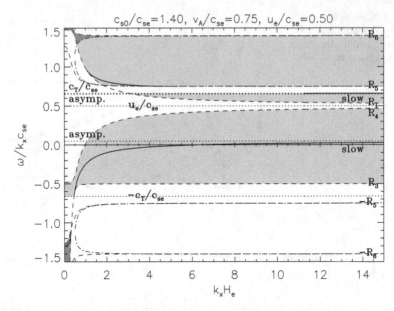

Fig. 5.6 Solution of the dispersion relation for MAG with flow for specific values; from Erdelyi et al. (1999)

where

$$m_0^2 = \frac{(k_x^2 V_A^2 - \omega^2)(k_x^2 c_s^2 - \omega^2)}{(c_s^2 + V_A^2)(k_x^2 c_T^2 - \omega^2)}$$

$$m_e^2 = \frac{k_x^2 c_s^2 - \Omega^2}{c_s^2}.$$

Equation (5.60) describes the parallel propagation of surface waves at a single magnetic interface in a gravitationally stratified atmosphere with the assumption of a constant Alfvén speed in the magnetic region and a constant homogeneous flow in the nonmagnetic region (see Figs. 5.6 and 5.7).

5.3 Waves in a Magnetic Slab

It is known that magnetic fields introduce structure (inhomogeneity) in an otherwise uniform medium, which will affect the wave propagation. One of the most sought-after structuring is that of an isolated magnetic slab. The disturbances outside the slab are ignored. To start with, the effect of gravity has been ignored (Roberts 1981). The field can support both body and surface waves. The existence and nature of these waves depend upon the relative magnitudes of the sound speed and the Alfvén speed inside the slab and that of the sound speed outside the slab. The slow mode, like the surface and body modes, always propagates, while the behavior of the fast

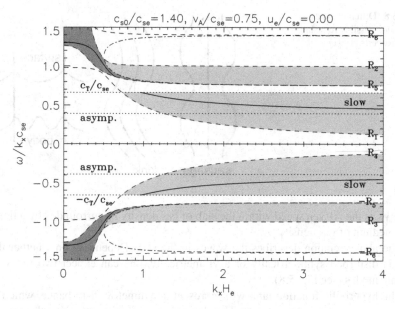

Fig. 5.7 Solution of the dispersion relation for MAG with flow for specific values; from Varga and Erdelyi (2001)

mode depends critically on the nature of the sound speeds on either side of the slab. The case of a slender flux tube has also been investigated. In this section, we shall consider the behavior of waves in a magnetic slab. The equilibrium is assumed to be

$$\mathbf{B}_0 = B_0, \qquad |x| < x_0$$
$$\mathbf{B}_0 = 0, \qquad |x| > x_0,$$

which is a uniform slab of magnetic field, whose width is $2x_0$ and that is surrounded by field-free plasma. The wave equation is the same as that in Eq. (5.19), and its solution in different regions can be written as

$$v_x(x) = \alpha_e e^{-k_z x} \qquad x > x_0$$
$$= \alpha_0 \cosh(k_z x) + \beta_0 \sinh(k_z x) \qquad |x| < x_0$$
$$= \beta_e e^{k_z x} \qquad x < -x_0$$

for arbitrary constants α_e, β_e, α_0, and β_0. The main boundary condition in addition to the boundedness of the solution is that the velocity component and the total pressure are continuous at the boundary $x = \pm x_0$. Applying these conditions results in the dispersion relation

$$\frac{k_z^2 V_A^2}{\omega^2} = 1 + \left[\frac{\rho_e}{\rho_0}\right] \left[\tanh(k_z x_0), \coth(k_z x_0)\right]. \qquad (5.61)$$

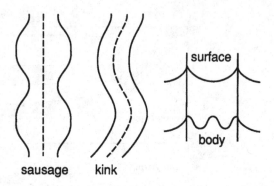

Here V_A is the Alfvén speed within the slab of gas density ρ_0 surrounded by a field-free medium of gas density ρ_e.

The above relation describes the waves in the slab. Depending on whether the slab is disturbed symmetrically or asymmetrically, one can expect two types of normal modes (see Fig. 5.8).

The hyperbolic function tanh will represent a symmetric disturbance, which is commonly called a sausage mode. The slab pulsates in such a way that the axis of symmetry remains undisturbed. For the asymmetric mode, the function coth will be the corresponding solution. The slab moves back and forth during the wave motion, and this mode is usually called the kink mode. The phase velocity of both these modes is less than the Alfvén speed of the slab.

It is easy to realize that unlike the hydromagnetic surface wave, those in the slab are dispersive waves; that is, the frequency of these modes is a function of the wavenumber and other physical parameters describing the system.

If we assume that the slab is sufficiently long with a small radius, it will be called a thin slab. The thin slab approximation corresponds to $k_z x_0 \ll 1$; that is, the long-wavelength disturbances propagate with a phase speed given by

$$\frac{\omega}{k_z} = V_A$$

for the sausage mode and

$$\frac{\omega}{k_z} = \left(\frac{\rho_0}{\rho_e}\right)^{1/2} (k_z x_0)^{1/2}$$

for the kink mode. The opposite situation would be a very wide slab wherein $k_z x_0 \gg 1$. In this case, the phase speed of both modes coincides and has the form

$$\frac{\omega}{k_z} = V_A \left[1 + \frac{\rho_e}{\rho_0}\right]^{-1/2},$$

clearly indicating that the behavior of these waves will be similar to an interfacial wave at a single interface that is field-free.

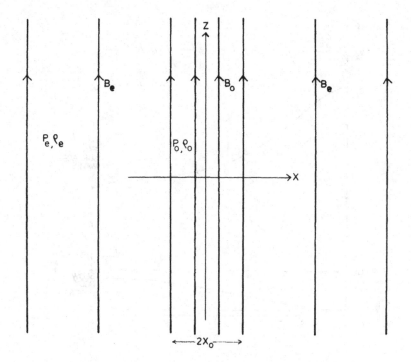

Fig. 5.9 Magnetic field in a structured slab; from Edwin and Roberts (1983)

Compressible case: The wave equation for the compressible slab geometry is the same as the wave equation (5.23). The analysis is similar to that of a single interface. The governing dispersion relation for magnetoacoustic slab waves is given by

$$(k_z^2 V_A^2 - \omega) m_e = \left(\frac{\rho_e}{\rho_0} \right) \omega^2 m_0 \, [\tanh, \coth] \, m_0 x_0. \qquad (5.62)$$

The restriction that m_0^2 should be positive in the case of a single interface need not be imposed in the slab. However, the condition that $m_e > 0$ still be valid as the solution of the above equation must be consistent with the condition $\omega^2 < k_z^2 c_e^2$. Waves with $m_0^2 > 0$ will continue to be called surface waves and those with $m_0^2 < 0$ the body waves (see Fig. 5.9).

The nomenclature of sausage and kink modes in the incompressible case will continue to hold for the compressible case, corresponding to the 'tanh' and 'coth' function, respectively. The solution of the wave equation inside the slab $(x < x_0)$ will be of the form

$$v_x = \sinh(m_0 x), \quad \cosh(m_0 x),$$

respectively, for the sausage and kink modes.

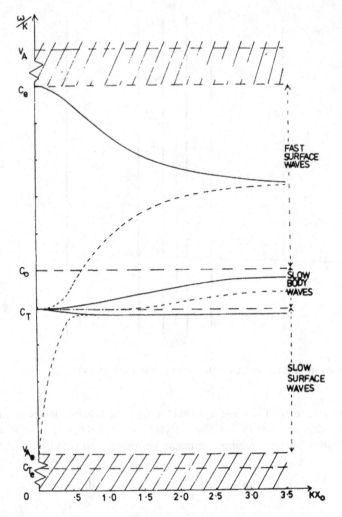

Fig. 5.10 The phase speed ω/k_z, plotted as a function of the nondimensional wavenumber $k_z x_0$; from Edwin and Roberts (1983)

The complicated transcendental equation for the dispersion relation prevents us from obtaining the expression for the phase speeds of the modes analytically, and we have to resort to solving the equation numerically. The numerical solution of the dispersion relation is given in Fig. 5.10, where the phase speed ω/k_z is plotted as a function of kx_0 for the case when $V_A > c_{se} > c_{s0}$.

In the incompressible limit where the sound speeds tend to infinity, so that m_0 and m_e tend to k_z, the resulting dispersion relation gives rise to two modes, which will be called the sausage surface mode and the kink surface mode. Also, in the wide slab limit ($k_z x_0 \gg 1$), replacing both $\tanh(m_0 x_0)$ and $\coth(m_0 x_0)$ by unity, one recovers the slow surface wave and also a fast surface wave.

The slow surface wave can easily be identified from Fig. 5.10 for the thin tube ($k_z x_0 \ll 1$). The slow sausage mode in the long-wavelength limit is given by

$$\omega \approx k_z c_T,$$

while the kink wave has the same approximate behavior as in the incompressible case.

Effect of flows inside the slab: In the above section, we discussed the behavior of waves in a slab geometry wherein the plasma was at rest, both inside and outside the slab. Now we shall concentrate on the behavior of the waves in a plasma slab that is moving uniformly with respect to the surrounding plasma. We assume that the plasma parameters, such as density, pressure, and magnetic field, are constant in each layer, but with a discontinuity at the interface separating the slab and the environment. As mentioned earlier, the wave equation may be written in terms of either the vertical velocity component or the total pressure. In what follows, we shall discuss the wave equation in terms of the total pressure for an incompressible fluid in a slab moving uniformly relative to the surrounding plasma. The dispersive characteristics of Alfvén surface waves (ASW) along a moving plasma surrounded by a stationary plasma has been studied Satya Narayanan and Somasundaram (1985). Also, the wave propagation in a magnetically structured compressible slab configuration has been investigated by Singh and Talwar (1993), allowing different field strengths inside and outside the slab, together with a general orientation of the field vectors relative to each other and the propagation vector. Singh and Talwar considered properties of body and surface waves for both symmetric and asymmetric modes of perturbation propagating along and normal to the slab field.

The properties of hydromagnetic surface waves along a plasma–plasma slab, when one of the fluids has a relative motion, has been studied as a function of the compressibility parameter c_S/V_A, with sound speed and Alfvén speed, respectively, by Satya Narayanan (1990). The properties of magnetosonic waves in a structured atmosphere with steady flows with applications to coronal and photospheric magnetic structures have been studied Nakariakov and Roberts (1995). In coronal loops, the appearance of backward slow body waves or the disappearance of slow body waves, depending on the direction of propagation, is possible if the flow speed exceeds the internal sound speed. Nakariakov et al. (1996) extended the above study to include waves trapped within solar wind flow tubes. According to them, the trapping is due to reflection of the waves from the tube boundary, which may correspond to either a jump in plasma density or magnetic field, or a jump in the steady flow velocity. They found that the phase and group speeds of these waves depend on the wave frequency and wavelength. They discuss two types of waves, the slow and the fast waves that get trapped. The phase speed of both waves is sub-Alfvénic.

The starting point for this discussion will be the wave equation:

$$\nabla^2 \hat{p} = 0, \tag{5.63}$$

where

$$\hat{p} = \bar{p} + \frac{B_{01} \cdot b}{4\pi\mu}.$$

Assuming wavelike perturbations that have a small amplitude, the solution for the wave equation can be written as

$$\hat{p}_1 = A_1 \sinh(kx),$$

where A_1 is an arbitrary constant. The solution for the pressure field for the stationary plasma surrounding the moving plasma is given by

$$\hat{p}_2 = B_1 e^{-kx},$$

where B_1 is an arbitrary constant. Applying the boundary conditions that the total pressure and the normal component of velocity are continuous, the dispersion relation may be simplified to yield an analytical expression for the nondimensional phase velocity as

$$\frac{\omega}{kV_A} = \frac{V \pm ([1+\eta\tanh(ka)][1+\beta^2\tanh(ka)] - \eta V^2\tanh(ka))^{1/2}}{1+\eta\tanh(ka)} \tag{5.64}$$

$$\frac{\omega}{kV_A} = \frac{V \pm ([1+\eta\coth(ka)][1+\beta^2\coth(ka)] - \eta V^2\coth(ka))^{1/2}}{1+\eta\coth(ka)}, \tag{5.65}$$

where $\beta = B_{02}/B_{01}$ and $\eta = \rho_{02}/\rho_{01}$ are the interface parameters, $V = U/V_A$ is a nondimensional velocity, V_A is the Alfvén velocity, and a is half the width of the moving plasma column. The first mode (5.64) is the symmetric mode and the second (5.65) is the asymmetric mode.

Special cases: In the limit $ka \to 0$, the above equations with $V = 0$ become

$$\frac{\omega}{kV_A} = 1$$

$$\frac{\omega}{kV_A} = \sqrt{\beta^2/\eta}$$

for the symmetric and asymmetric modes, respectively. In this case, the symmetric mode is independent of the interface parameters β and η, which is not the case for the asymmetric mode, as seen from the above expressions.

In the limit $ka \to \infty$, both $\tanh(ka)$ and $\coth(ka) \to 1$, so that Eqs. (5.64) and (5.65) reduce to

$$\frac{\omega}{kV_A} = \frac{V \pm [(1+\eta)(1+\beta^2) - \eta V^2]}{1+\eta}. \tag{5.66}$$

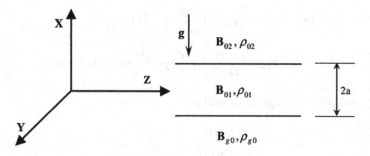

Fig. 5.11 The geometry; from Satya Narayanan et al. (2009)

The phase velocity of both modes coincides, unlike the case $ka \to 0$. For $V = 0$, the normalized phase speed reduces to

$$\frac{\omega}{kV_A} = \left[\frac{1+\beta^2}{1+\eta}\right]^{1/2}.$$

For the case $V = 0$ and $\beta = 0$, no flow and outside magnetic field being absent,

$$\frac{\omega}{kV_A} = \pm\left[\frac{1}{\left[1+\eta(\tanh, \coth)ka\right]}\right]^{1/2}$$

for the symmetric and asymmetric modes, respectively. For a plasma slab in vacuum, $\eta = 0$, so that

$$\frac{\omega}{kV_A} = V \pm \left[\left[1+\beta^2(\tanh, \coth)ka\right]\right].$$

Effect of flows and gravity: It is also well known that gravity waves play an important role in studying the coupling of lower and upper solar atmospheric regions and are therefore of tremendous interdisciplinary interest (Satya Narayanan et al. 2004a). The gravity waves in the Sun may be divided into two types, namely, (1) the internal gravity waves, which are confined to the solar interior, and (2) the atmospheric gravity waves, which are related to the photosphere and chromosphere, and may be further beyond. In general, the observation of gravity mode oscillations of the Sun would provide a wealth of information about the energy-generating regions, which is poorly probed by the p-mode oscillations. The effect of uniform flows and gravity will be discussed here. We have assumed a slab of width "2a," having different densities in each layer, and uniform flow in the plasma slab, with the gravity acting downward (see Fig. 5.11). For the sake of brevity, we shall skip the details on the derivation of the dispersion relation. For the sake of simplicity, we have adopted the plasma β to be very small. The relation that governs the MHD waves, coupling the slab and its environment, is given by

$$\left[\varepsilon_1\varepsilon_2 + \varepsilon_1\varepsilon_g + g\varepsilon_1\frac{(\rho_g - \rho_2)}{\omega}\right] + \left[\varepsilon_1^2 + \varepsilon_2\right.$$

$$\left.\varepsilon_g + \varepsilon_2 g\frac{(\rho_g - \rho_1)}{\Omega}\varepsilon_g g\frac{(\rho_1 - \rho_2)}{\Omega}\right]\tanh(2ka) = 0. \qquad (5.67)$$

The epsilons and omega are defined as

$$\Omega = \omega - kU_0$$

$$\varepsilon_1 = \frac{(\rho_1\Omega)}{k(v_{ph} - \hat{U})^2}[1 - (v_{ph} - \hat{U})^2]$$

$$\varepsilon_2 = \frac{(\rho_1\omega)}{kv_{ph}^2}[\beta_1^2 - v_{ph}^2\eta_1]$$

$$\varepsilon_g = \frac{(\rho_1 g)}{kv_{ph}^2}[\beta^2 - v_{ph}^2\eta].$$

Introducing the nondimensional quantities

$$v_{ph} = \frac{\omega}{kv_{A1}}, \hat{U} = \frac{U_0}{v_{A1}}, \beta = \frac{B_{0g}}{B_{01}}, \beta_1 = \frac{B_{02}}{B_{01}},$$

$$\eta = \frac{\rho_{0g}}{\rho_{01}}, \eta_1 = \frac{\rho_{02}}{\rho_{01}} G = \frac{g}{kv_{A1}^2},$$

the normalized dispersion relation can be simplified to yield

$$[1 - (v_{ph} - \hat{U})^2][\beta_1^2 - v_{ph}^2\eta_1) + (\beta^2 - v_{ph}^2\eta) + G(\eta - \eta_1)]$$

$$\left[\frac{v_{ph}}{(v_{ph} - \hat{U})}(1 - (v_{ph} - \hat{U})^2)^2 + \frac{(v_{ph} - \hat{U})}{v_{ph}}(\beta_1^2 - v_{ph}^2\eta_1)(\beta^2 - v_{ph}^2)\right.$$

$$\left. + (\beta_1^2 - v_{ph}^2\eta_1)G(\eta - 1) + (\beta^2 - v_{ph}^2\eta)G(1 - \eta_1)\right]\tanh(2ka) = 0. \qquad (5.68)$$

The dispersion relation, which is solved numerically for the phase speed, is plotted as function of the dimensionless wavenumber for various values of the interface parameters $\beta^2 = 0.5$, $\beta_1^2 = 1.5$, $\eta = 1.8$, $\eta_1 = 1.2$ and different values of the nondimensionalized G and $\hat{U}$, as shown in Fig. 5.12. For the fully compressible case, both the fast Alfvén gravity surface wave and the slow Alfvén gravity surface wave are present. However, since the plasma beta is small, the slow mode disappears. For increasing values of G, the normalized phase speed of these waves increases as a function of ka. This situation is true when there is no flow. However, when flow is introduced, the phase speed is significantly reduced. This shows that flows tend to dampen the phase speed of the fast Alfvén gravity surface waves. It is

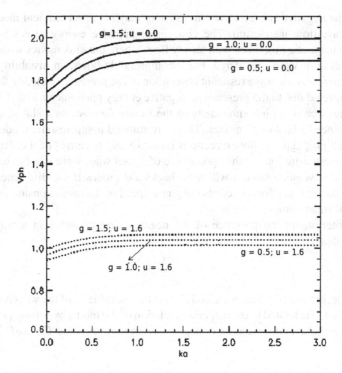

Fig. 5.12 The normalized phase velocity as a function of the nondimensional wavenumber; from Satya Narayanan et al. (2009)

interesting to note that the variation in the phase speed is significant only for $ka \approx 1$, while for $ka > 1$, the speed asymptotically approaches the phase speed of the body wave.

Negative energy waves: Different plasma structures show the presence of steady flows of the matter, which in general is directed along the direction of the magnetic field. In the Sun, there are downflows along the photospheric magnetic flux tubes due to the presence of granules and supergranules, spicules in the chromosphere. The presence of homogeneous steady flows generally leads to the Doppler shift in the wave frequencies. Whenever there is a shift in the steady flow, there arises what are called the Kelvin–Helmholtz instabilities, which we will discuss in Chap. 6. It has been found that the presence of the transverse shift in the steady flows can lead to the appearance of backward waves. Slow steady-flow velocities can change the dispersive characteristics of the MHD modes in magnetic structures.

An interesting observation about the general theory of waves is that backward waves (retarded) may have negative energy (Ruyotova 1988; Joarder et al. 1997). The implication is that these waves tend to grow when the total energy of the system is decreased. The amplification of these waves can occur when there is dissipation in the system. This does not lead to any contradiction to the law of conservation of energy. The negative energy waves in a system may act as an efficient mechanism for the wave-flow interaction. The concept of negative energy waves has been studied in magnetic flux tubes with steady flow, where the kink modes with a long-wavelength limit were considered. For incompressible flows, an instability due to surface magnetosonic wave resonant absorption in the presence of steady flows has been interpreted due to the presence of negative energy phenomena.

The presence of an inhomogeneity in the steady flow across a slab may cause an appearance of backward modes. There are trapped magnetosonic modes in the slab, which propagate in both directions (positive and negative) of the axis of the slab. The main criterion for the appearance of a backward wave can be defined as follows: A waveguide system will have backward modes if the difference in the speeds of the external flows exceeds the phase speed of the mode considered, in the absence of steady flows.

The criterion for the presence of the negative energy waves in a waveguide system is that

$$C = \omega \frac{\partial D}{\partial \omega} < 0, \tag{5.69}$$

where $D(\omega, k_0)$ is a function evaluated at a particular value k_0 of the wavenumber k. The function D is related to the dispersion relation of the mode, by setting $D = 0$. An alternate formula for the presence of negative energy waves is that $C = k \partial D / \partial k < 0$.

5.4   Waves in Cylindrical Geometries

The appearance of many magnetic structures that have a cylindrical flux tube shape in low-β plasmas in the magnetosphere and corona , in particular in the Sun, encourages one to study the waves in cylindrical geometries. The properties of Alfvén surface waves along a cylindrical plasma column surrounded by vacuum or another plasma column has been discussed by Uberoi and Somasundaram (1982a). Both the symmetric (m=0) and asymmetric ($m = \pm 1$) modes are found to be dispersive in nature. The interfacial symmetric mode propagates in a certain frequency window ($\omega_{V_{A1}}, \omega_{V_{As}}$), where $\omega_{V_{As}}$ is the Alfvén surface wave frequency along the interface of the two semiinfinite media. The symmetric mode can be converted into a forward wave from a backward wave for a critical wavenumber, depending on the choice of the bulk Alfvén speeds on either side of the media. This study was extended to include moving plasma columns in cylindrical geometry by Somasundaram and Satya Narayanan (1987). The effect of compressibility on the nature of these waves for a moving plasma in cylindrical geometry was studied by Satya Narayanan (1991).

The equations are similar to those of the slab geometries, except that the solutions of the wave equations are not in terms of hypergeometric functions, but in terms of the cylinder functions (Bessel functions).

Magnetic flux tubes act as wave guides, allowing waves to propagate without spatial attenuation. So they act as a good communication channel between one

Fig. 5.13 The cylindrical geometry with uniform flow inside the tube; from Satya Narayanan et al. (2004)

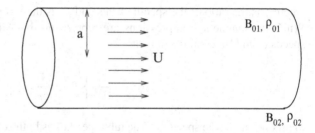

region of the plasma and another. In much the same way, the loops of magnetic field in the coronal atmosphere may act as communication channels. The flux tubes support a variety of MHD waves, including the Alfvén waves, the magnetosonic waves, and so on. Sunspots support a variety of wave phenomena that may be interpreted in terms of MHD waves. The presence of MHD waves in the solar corona was just a theoretical suggestion until recently.

The evidence for the occurrence of magnetic flux tubes in astrophysical phenomena is increasing all the time. Flux tubes (or flux ropes) are believed to occur in the jets in some of the extragalactic sources, in the magnetospheres of some planets, and in the Sun. Indeed, in the case of the Sun, almost all of the emerging flux through the photospheric surface is found to occur in concentrated forms, ranging in scale from the visible sunspot to the very small ≈ 1" intense flux tubes. Thus, the source of coronal or solar wind magnetism is to be found in concentrated roots of magnetic field emerging from the deep interior. First, we note that a tube is a wave guide. It permits waves to propagate without spatial attenuation. Thus, a tube is likely to provide a good communication channel between one region of a plasma and another, perhaps providing a connection between an energy source and an energy sink. For example, photospheric flux tubes connect the convection zone—an ample source of energy, especially in granules—with the chromosphere and the corona. Second, we observe that a magnetic flux tube is an elastic object (and so an elastic, not rigid, wave guide) and as such is likely to respond to sudden changes by guiding waves. Sunspots are known to support a wide variety of wave phenomena. From an observational point of view, the clearest evidence for flux tubes as magnetically distinct structures exists in the solar photosphere.

Different Types of Modes in Cylindrical Geometry

We consider the modes of oscillations of an isolated magnetic flux tube of strength B_0 embedded in a field free of pressure p_e and density ρ_e (see Fig. 5.13). If such a tube is subjected to a sudden twisting motion, it will respond by producing torsional oscillations of the tube. Such torsional oscillations may propagate as an Alfvén wave, for a tube of density ρ_0 and magnetic permeability μ. Compression of the gas is characterized by the sound speed:

$$c_{s0} = \frac{\gamma p_0}{\rho_0}^{1/2},$$

where γ is the ratio of the specific heats and p_0 is the gas pressure inside the tube. While the magnetic compressions are represented by Alfvén speed V_A, the two speeds c_s and V_A combine to form

$$\left\{ \frac{1}{c_s^2} + \frac{1}{V_A^2} \right\}^{1/2} = \frac{1}{c_T}$$

to produce the basic speed c_T. The tube speed c_T is both sub-Alfvénic and subsonic and therefore relates to the slow magnetoacoustic wave. A symmetric squeezing of the tube may produce the sausage mode. Here, both gas and magnetic field within the tube are expanding and contracting in the motion.

Asymmetric disturbances of the tube produces kink modes. This is similar to the waves on an elastic string, producing a propagation speed of $(T/\rho)^{1/2}$. The tension T in the string here is clearly due to the magnetic tension, so $T = B_0^2/\mu_0$, and the density ρ is taken to be the sum of the gas density within the tube (ρ_0) and in the environment (ρ_e); that is, $\rho = \rho_0 + \rho_e$.

The kink speed becomes

$$c_k = \frac{B_0}{[\mu_0(\rho_0 + \rho_e)]^{1/2}} = \frac{B_0}{[\mu_0\rho_0 + \mu_0\rho_e]^{1/2}} = \frac{1}{(\mu_0\rho_0)^{1/2}/B_0 + (\mu_0\rho_e)^{1/2}/B_0}.$$

$$(5.70)$$

We have $V_A = B_0/(\mu_0\rho_0)^{1/2}$. Thus,

$$c_k = V_A \left\{ \frac{\rho_0}{\rho_0 + \rho_e} \right\}^{1/2}.$$

The kink speed c_k is sub-Alfvénic but not necessarily subsonic. Additionally, the radial behavior of the amplitude within $r \leq a$ may be either oscillating or decaying. Oscillating modes are classified as body waves, while decaying or evanescent modes are surface waves. A schematic diagram of these modes has already been presented.

We consider a uniform cylinder of magnetic field $B_0\hat{z}$, confined to a radius of a surrounded by a magnetic field $B_e\hat{z}$. The gas pressure and density within the cylinder are p_0 and ρ_0 and outside it are p_e and ρ_e.

Pressure balance implies

$$p_0 + \frac{B_0^2}{2\mu} = p_e + \frac{B_e^2}{2\mu},$$

where μ is the magnetic permeability.

The sound and Alfvén speeds inside and outside the cylinder are given by

$$c_{s0} = (\gamma p_0/\rho_0)^{1/2} \quad \Rightarrow \quad p_0 = \rho_0 c_0^2/\gamma$$

$$V_{A0} = \frac{B_0}{(\mu\rho_0)^{1/2}} \quad \Rightarrow \quad B_0 = V_{A0}(\mu\rho_0)^{1/2}$$

$$c_e = (\gamma p_e / \rho_e)^{1/2} \quad \Rightarrow \quad p_e = \rho_e c_e^2 / \gamma$$

$$V_{Ae} = \frac{B_e}{(\mu \rho_e)^{1/2}} \quad \Rightarrow \quad B_e = V_{Ae}(\mu \rho_e)^{1/2},$$

where γ is the ratio of specific heats.

Substituting the above relations for magnetic fields and pressure, we get the relation connecting the densities ρ_0 and ρ_e; that is,

$$\frac{\rho_0}{\rho_e} = \frac{2c_0^2 + \gamma V_{A0}^2}{2c_e^2 + \gamma V_{Ae}^2}.$$

Linear perturbations about this equilibrium lead to two equations (Erdelyi 2008):

$$\frac{\partial^2}{\partial t^2}\left(\frac{\partial^2}{\partial t^2} - (c_0^2 + V_A^2)\nabla^2\right)\triangle + c_{s0}^2 V_A^2 \frac{\partial^2}{\partial z^2}\nabla^2 \triangle = 0 \qquad (5.71)$$

$$\left(\frac{\partial^2}{\partial t^2} - V_A^2\frac{\partial^2}{\partial z^2}\right)\Gamma = 0, \qquad (5.72)$$

where ∇^2 is the Laplacian operator in cylindrical coordinates (r, θ, z);

$$\nabla^2 = \frac{\partial^2}{\partial r^2} + \frac{1}{r}\frac{\partial}{\partial r} + \frac{1}{r^2}\frac{\partial^2}{\partial \theta^2} + \frac{\partial^2}{\partial z^2}$$

and

$$\Gamma = \hat{z} \cdot \text{Curl}V = \frac{1}{r}\frac{\partial}{\partial r}(rV_\theta) - \frac{1}{r}\frac{\partial v_r}{\partial \theta},$$

where $V = (v_r, v_\theta, v_z)$.

If $\triangle = R(r)\exp[i(\omega t + n\theta + kz)]$, then the equations imply that $R(r)$ satisfies the Bessel's equation given by

$$\frac{d^2 R}{dr^2} + \frac{1}{r}\frac{dR}{dr} - \left(m_0^2 + \frac{n^2}{r^2}\right)R = 0, \qquad (5.73)$$

where

$$m_0^2 = \frac{(k^2 c_0^2 - \omega^2)(k^2 V_A^2 - \omega^2)}{(c_0^2 + V_A^2)(k^2 c_T^2 - \omega^2)}. \qquad (5.74)$$

For a solution bounded on the axis ($r = 0$) of the cylinder

$$R(r) = A_0 I_n(m_0 r), \qquad m_0^2 > 0$$

$$= A_0 J_n(m_0 r), \qquad m_0^2 < 0,$$

A_0 is a constant. I_n and J_n are Bessel functions of order n.

In the external region, with no propagation of energy away from R, the solution is given by

$$R(r) = A_1 K_n(m_e r), r > a,$$

where

$$m_e = \frac{(k^2 c_e^2 - \omega^2)(k^2 V_{Ae}^2 - \omega^2)}{(c_e^2 + V_{Ae}^2)(k^2 c_{Te}^2 - \omega^2)},$$

where

$$c_{Te}^2 = \frac{c_e^2 V_{Ae}^2}{c_e^2 + V_{Ae}^2}.$$

Here m is taken to be positive.

Continuity of radial velocity component v_r and the total pressure across the cylinder (gas + magnetic) boundary ($r = a$) yields the required dispersion relations:

$$\rho_0(k^2 V_{A0}^2 - \omega^2) m_e \frac{K_n'(m_e a)}{K_n(m_e a)} = \rho_e (k^2 V_{Ae}^2 - \omega^2) m_0 \frac{I_n'(m_0 a)}{I_n(m_0 a)} \tag{5.75}$$

for surface waves $m_0^2 > 0$ and

$$\rho_0(k^2 V_{A0}^2 - \omega^2) m_e \frac{K_n'(m_e a)}{K_n(m_e a)} = \rho_e(k^2 V_{Ae}^2 - \omega^2) n_0 \frac{J_n'(n_0 a)}{J_n(n_0 a)} \tag{5.76}$$

for body waves $m_0^2 = -n_0^2 < 0$.

Here $n = 0$ represents the cylindrically symmetric sausage mode, while $n = 1$ represents the asymmetric kink mode (see Fig. 5.14). The effect of uniform flows on the characteristics of waves in flux tubes has been studied.

Slender Flux Tube Equations

Slender magnetic flux tubes are typically tubes wherein the vertical motions slowly diverge with height z and have a radius much smaller than the pressure scale height. The governing equations are the equation of continuity, vertical momentum, transverse momentum, and isentropic energy, given by

$$\frac{\partial}{\partial t}\rho A + \frac{\partial}{\partial z}\rho v A = 0 \tag{5.77}$$

$$\rho\left[\frac{\partial v}{\partial t} + v\frac{\partial v}{\partial z}\right] = -\frac{\partial p}{\partial z} - \rho g \tag{5.78}$$

$$p + \frac{B^2}{2\mu} = p_e \tag{5.79}$$

$$\frac{\partial p}{\partial t} + v\frac{\partial p}{\partial z} = \frac{\gamma p}{\rho}\left[\frac{\partial \rho}{\partial t} + v\frac{\partial \rho}{\partial z}\right] \tag{5.80}$$

$$BA = \text{constant}. \tag{5.81}$$

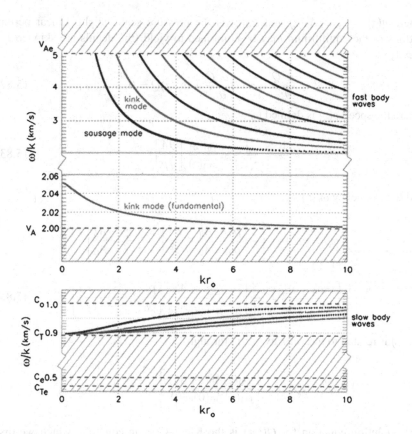

Fig. 5.14 Solution of the dispersion relation for the cylindrical geometry; from Erdelyi (2008)

The above equations govern the nonlinear behavior of longitudinal, isentropic motions $v(z,t)$ of a gas of density $\rho(z,t)$ and pressure $p(z,t)$ confined within an elastic tube of cross-sectional area $A(z,t)$. The derivation of the equations, as an expansion about the axis of the tube, has been reported. The case of the incompressible fluid has been considered. Also considered were the cases of isothermal and nonisothermal effects on the thin flux tube equations. By setting $v = 0$ and $\partial/\partial t = 0$, the equilibrium will be recovered. Assuming that the external medium is in hydrostatic equilibrium, with the same temperature and scale height inside and outside the tube, the thin flux tube equations reduce to

$$p_0(z)p_0(0)e^{-n(z)} \quad \rho_0(z) = \rho_0\frac{\Lambda(0)}{\Lambda(z)}e^{-n(z)}$$

$$A_0(z) = A_0(0)e^{(1/z)n(z)},$$

where $n(z) = \int_0^z dz/\Lambda_0(z)$ and $\Lambda_0(z)$ is the pressure scale height. Linear perturbations of the equilibrium may be combined to yield (the details are skipped for brevity)

$$\frac{\partial^2 Q}{\partial t^2} - c^2(z)\frac{\partial^2 Q}{\partial z^2} + \omega_v^2(z)Q = 0, \qquad (5.82)$$

where the speed $c(z)$ is defined by

$$\frac{1}{c^2} = \frac{1}{c_s^2(z)} + \frac{\rho_0(z)}{\Lambda_0(z)}\left[\frac{\partial A}{\partial p}\right]_{p=0}, \qquad (5.83)$$

and the frequency $\omega_v(z)$ by

$$\omega_v^2 = \omega_g^2 + c^2 \left[\frac{1}{2}\left[\frac{\rho_0'}{\rho_0} + \frac{A_0'}{A_0} + \frac{c'^2}{c^2}\right]' + \left[\frac{g}{c_s^2} - \frac{A_0'}{A_0}\right]' \right.$$

$$\left. \frac{1}{4}\left[\frac{rho_0'}{\rho_0} + \frac{A_0'}{A_0} + \frac{c'^2}{c^2}\right]^2 + \left[\frac{g}{c_s^2} - \frac{A_0'}{A_0}\right]\left[\frac{\rho_0'}{\rho_0} + \frac{c'^2}{c^2} + \frac{g}{c_s^2}\right] \right]. \qquad (5.84)$$

$Q(z,t)$ is related to the velocity $v(z,t)$ through

$$Q(z,t) = \left[\frac{\rho_0(z)\Lambda_0(z)c^2(z)}{\rho_0(0)\Lambda_0(0)c^2(0)}\right]^{1/2} v(z,t). \qquad (5.85)$$

The evolution equation for $Q(z,t)$ is the Klein–Gordon equation, which we discussed for the symmetric sausage mode.

The analysis for the kink mode is slightly different. The linear transverse motions $v_\perp$ are governed by the equation

$$\frac{\partial^2 v_\perp}{\partial t^2} = g\left[\frac{\rho_0 - \rho_e}{\rho_0 + \rho_e}\right]\frac{\partial v_\perp}{\partial z} + \left[\frac{\rho_0}{\rho_0 + \rho_e}\right]V_A^2\frac{\partial^2 v_\perp}{\partial z^2}. \qquad (5.86)$$

The first term on the right-hand side represents the buoyancy effects on the isolated flux tube, while the second term deals with the restoring force of the magnetic tension.

To get the Klein–Gordon equation for the kink mode, we write

$$v_\perp(z,t) = e^{z/4\Lambda_0}Q(z,t). \qquad (5.87)$$

5.5 Waves in Untwisted and Twisted Tubes

Oscillations in annular magnetic cylinders: Here, we shall consider a flux tube consisting of a central core surrounded by a shell or annulus layer, embedded in a uniform magnetic field. To begin with, we shall deal with an incompressible fluid wherein the phase speed of the slow and Alfvén modes becomes indistinguishable (Erdelyi and Carter 2006; Carter and Erdelyi 2007, 2008). The fast modes are neglected. The longitudinal magnetic field in each of the regions will be as follows:

$$\mathbf{B} = B_i = (0,0,B_i) \quad r < a$$
$$\mathbf{B} = B_0 = (0,0,B_0) \quad a \leq r \leq R$$
$$\mathbf{B} = B_e = (0,0,B_e) \quad r > R,$$

where B_i, B_0, B_e are constant. The densities at the core, annulus, and external regions will be ρ_i, ρ_0, and ρ_e, respectively. A similar expression for the pressure distribution will be assumed. The pressure balance conditions at the boundaries $r = a$ and $r = R$ are given by

$$p_i + \frac{B_i}{2\mu} = p_0 + \frac{B_0}{2\mu}$$
$$p_0 + \frac{B_0}{2\mu} = p_e + \frac{B_e}{2\mu}.$$

The distribution of the magnetic field in the annulus is shown in Fig. 5.15.

Assuming linear perturbations of the ideal MHD equations about the equilibrium and Fourier-transforming the total Lagrangian pressure p_T and normal component of Lagrangian displacement ξ_r,

$$(p_T, \xi_r) \sim (\hat{p}_T(r), \hat{\xi}_r(r))e^{i(m\theta + k_z z - \omega t)} \tag{5.88}$$

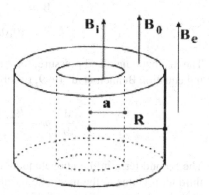

Fig. 5.15 Distribution of the magnetic field in the annulus; from Carter and Erdelyi (2007)

and omitting the hat of the Fourier-decomposed perturbations for the sake of brevity, the total pressure satisfies the Bessel equation as given below:

$$\frac{d^2 p_T}{dr^2} + \frac{1}{r}\frac{dp_T}{dr} - (k_z^2 + \frac{m^2}{r^2})p_T = 0, \tag{5.89}$$

where m is the azimuthal wavenumber. The dispersion relation, after substituting the boundary conditions at the boundaries $r = a$ and $r = R$ and some algebra, reduces to

$$\frac{Q_0^i K_m'(k_z a) - (I_m'(k_z a)K_m(k_z a)/I_m(k_z a))}{I_m'(k_z a)(Q_0^i - 1)}$$

$$= \frac{K_m'(k_z R)(Q_0^e - 1)}{Q_0^e I_m'(k_z R) - (K_m'(k_z R)I_m(k_z R)/K_m(k_z R))}, \tag{5.90}$$

wherein

$$Q_0^i = \frac{\rho_i}{\rho_0}\frac{(\omega^2 - \omega_{Ai}^2)}{(\omega^2 - \omega_{A0}^2)} \tag{5.91}$$

$$Q_0^e = \frac{\rho_e}{\rho_0}\frac{(\omega^2 - \omega_{Ae}^2)}{(\omega^2 - \omega_{A0}^2)}. \tag{5.92}$$

There are two modes (surface) to the above dispersion relation for each of the sausage and kink modes, respectively, for the annulus–core model. These modes propagate along the two natural surfaces of the system, namely, at $r = a$ and $r = R$. The phase speeds are modified due to the annulus in comparison with a single straight tube, where the modes depend not only on the Alfvén speed, but also on the ratio (a/R) of the core and annulus radii.

Magnetically twisted cylindrical tube: Consider a flux tube embedded in a straight magnetic field (see Fig. 5.16) given by

$$\mathbf{B} = (0, Ar, B_0) \qquad r < a$$

$$\mathbf{B} = (0, 0, B_e) \qquad r > a.$$

The magnetic field and pressure, for the cylindrical equilibrium, satisfy the following equation Bennett et al. 1999; Erdelyi and Fedun 2006, 2007; 2010:

$$\frac{d}{dr}\left[p_0 + \frac{B_{0\phi}^2 + B_{0z}^2}{2\mu}\right] + \frac{B_{0\phi}^2}{\mu r} = 0. \tag{5.93}$$

The second term in the above equation represents the magnetic pressure, and the third term is due to the magnetic tension, derived due to the azimuthal component of the equilibrium magnetic field. Again, here we discuss only the incompressible

Fig. 5.16 Distribution of the
magnetic field in the twisted
loop; from Erdelyi and Fedun
(2007)

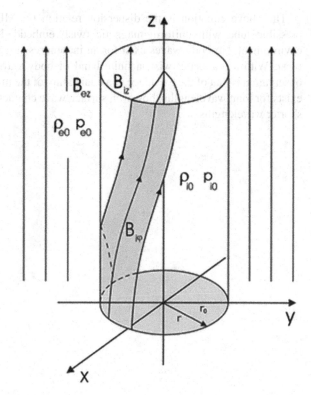

case. We seek a bounded solution at $r = 0$ and $r \rightarrow \infty$. Applying the usual boundary
conditions results in the following dispersion relation:

$$
\frac{(\omega^2 - \omega_{A0}^2) \frac{m_0 a I'_m(m_0 a)}{I_m(m_0 a)} - 2m\omega_{A0} \frac{A}{\sqrt{\mu\rho_0}}}{(\omega^2 - \omega_{A0}^2)^2 - 4\omega_{A0}^2 \frac{A^2}{\mu\rho_0}}
$$

$$
= \frac{\frac{|k_z| a K'_m(|k_z|a)}{K_m(|k_z|a)}}{\frac{\rho_e}{\rho_0}(\omega^2 - \omega_{Ae}^2) + \frac{A^2}{\mu\rho_0} \frac{|k_z| a K'_m(|k_z|a)}{K_m(|k_z|a)}}. \tag{5.94}
$$

The dash in the above equation denotes derivative with respect to the argument of
the Bessel function, and the frequencies are defined as

$$
\omega_{A0} = \frac{1}{\sqrt{\mu\rho_0}}(mA + k_z B_0), \qquad \omega_{Ae} = \frac{k_z B_e}{\sqrt{\mu\rho_e}} \tag{5.95}
$$

$$
m_0^2 = k_z^2 \left[1 - \frac{4A^2 \omega_{A0}^2}{\mu\rho_0(\omega^2 - \omega_{A0}^2)^2} \right]. \tag{5.96}
$$

The above equation is the dispersion relation for MHD waves in an incompressible tube with uniform magnetic twist, embedded in a straight magnetic environment. No body waves exist for an incompressible fluid without a magnetic twist. With a magnetic twist, a finite band of body waves arises. An interesting observation is that of the existence of a dual nature of the mode wherein a body wave exists for long wavelengths; however, surface wave characteristics are displayed for shorter wavelengths.

Chapter 6
Instabilities

6.1 Introduction

In Chap. 3, we discussed some exact solutions of MHD, force-free fields, and other types of solutions in hydrodynamics. An important aspect of these solutions, irrespective of whether one deals with hydrodynamics, MHD, or even plasmas, is the notion of stability. One of the famous equations that deal with MHD equilibrium is the Grad–Shafaranov equation, whose solution provides the state of a static MHD equilibrium. The important question that may be asked is whether this equilibrium is stable.

As an example of describing the notion of stability, let's consider a plasma ball, located at the bottom of a valley or on the top of a hill (both have dimensions comparable to that of the ball). If the ball is at the bottom of the valley, it will possess a minimum of the potential energy, and a slight lateral displacement will result in a restoring force, which pushes the ball back to its original position. The ball tends to overshoot from its equilibrium position and oscillates about the minimum with a constant amplitude (as the energy is conserved). On the other hand, if the ball is located on the top of the hill, a slight lateral displacement forces the ball to be pushed to the side, which results in an increase of the velocity. This perturbation does not restore the ball to its original position but is in the direction of the original displacement. Thus, there will not be an oscillation in the velocity.

Consider a simple harmonic system as follows:

$$m\frac{d^2x}{dt^2} = \pm\kappa x,\tag{6.1}$$

where κ is assumed to be positive. The plus and minus signs in the above equation should be chosen depending on whether the ball is on the hill or in the valley, respectively. It is trivial to realize that the above equation has a solution $x \sim \exp(-i\omega t)$, where $\omega = \pm\sqrt{\kappa/m}$ for the valley and $\omega = i \pm \sqrt{\kappa/m}$ for the hill. The case when $\omega = +i\sqrt{\kappa/m}$ results in an increase in the amplitude "x" as a

A. Satya Narayanan, *An Introduction to Waves and Oscillations in the Sun*, Astronomy and Astrophysics Library, DOI 10.1007/978-1-4614-4400-8_6, © Springer Science+Business Media New York 2013 function of time. This solution is unstable and corresponds to the ball accelerating down the hill when it is perturbed from its initial equilibrium solution. The same argument may be extended to two dimensions. In this case, for instability an absolute minimum in both directions would be required. A saddle-point potential would serve such a purpose, for in this case, the ball would always roll down from the saddle point. Using a similar argument, one can say that for a multidimensional system, the stability is assured if the equilibrium potential energy corresponds to an absolute minimum with respect to all possible displacements.

The task of determining MHD stability is similar to having a ball in a multidimensional system. If the potential energy of the system increases for any allowed perturbation of the system, then the system is stable. However, if even one allowed perturbation that decreases the potential energy of the system exists, then the system will become unstable.

There are several instabilities in the literature. However, in this book, we shall restrict ourselves to only four, namely, the Rayleigh–Taylor instability, Kelvin–Helmholtz instability, parametric instability, and Parker instability.

In what follows, at the end of each of the sections, we shall give one or more applications to the Sun of the instabilities discussed below.

6.2   Rayleigh–Taylor Instability

To begin, let's discuss Rayleigh–Taylor (RT) instability initially for hydrodynamics and then move on to the discussion on MHD. One can think of a situation wherein a heavy block is kept on a lighter block and achieves equilibrium. However, if one asks whether such a configuration is stable, the answer will be negative. In hydrodynamics, a situation similar to that of the blocks is that of a heavier fluid supported by a lighter fluid. A small rippling motion would disturb the equilibrium and make it unstable. The ripples are unstable because they effectively interchange volume elements of heavy fluid with equivalent volume elements of lighter fluid. Each volume element of interchanged heavy fluid originally had its center of mass a distance $\triangle$ above the interface, while each volume element of interchanged light fluid originally had its center of mass a distance $\triangle$ below the interface. The potential energy of a mass m at a height h is given by mgh, where g is the gravitational field. If we calculate the respective changes in the potential energy of the lighter and heavier fluids, we get

$$\delta W_h = -2\rho_h V \triangle g, \quad \delta W_l = +2\rho_l V \triangle g. \tag{6.2}$$

Here, V is the volume of the interchanged fluid elements and ρ_h and ρ_l are the densities (mass) of the heavier and lighter fluids, respectively. The net change in the potential energy of the total system is

$$\delta W = -2(\rho_h - \rho_l)V \triangle g, \tag{6.3}$$

which is less than zero. So the system lowers its potential energy by forming ripples. This is similar to the ball rolling from the top of the hill.

A classic example of this instability is that of the inverted glass of water. The heavy fluid in this case is the water and the lighter fluid is the air. This system is stable when a piece of cardboard is located at the interface between the water and the air. However, when the cardboard is removed, the system becomes unstable and the water starts to fall out. The cardboard prevents the ripple interchange from happening. The system remains stable when the ripples are prevented, for the atmospheric pressure is adequate to support the inverted water. The cardboard places a constraint on the system, with a boundary condition that prevents ripple formation. However, when the cardboard is removed, there is no longer any constraint against ripple formation. Thus, the ripples tend to grow into large amplitudes with the result that water falls. This is a typical case of an unstable equilibrium. The reader may look into the following books on topics related to Rayleigh–Taylor (RT) and Kelvin–Helmholtz (KH) instabilities (Chandrasekhar 1961; Chen 1977; Bellan 2006; Dendy 1990).

Let's work on the above argument in hydrodynamics with its usual equations of motion, so that we can get a quantitative idea on the stability of such a system. Consider the stability of a heavier fluid such as water, supported by a light fluid, which is air, such that there is no constraint at the interface. Assume the vertical in the y-direction so that gravity acts in the negative direction. Let's also assume that $\rho_l \ll \rho_h$, so that the mass of the lighter fluid can be ignored. To begin, the water and air are assumed to be incompressible, no variation in the density. The continuity equation is

$$\frac{\partial \rho}{\partial t} + \mathbf{v} \cdot \nabla \rho + \rho \nabla \cdot \mathbf{v} = 0. \tag{6.4}$$

For the incompressible fluid, it reduces to

$$\nabla \cdot \mathbf{v} = 0. \tag{6.5}$$

The linearized continuity equation in the water reduces to

$$\frac{\partial \rho_1}{\partial t} + \mathbf{v}_1 \cdot \nabla \rho_0 = 0, \tag{6.6}$$

and the linearized equation of motion in the water is

$$\rho_0 \frac{\partial \mathbf{v}_1}{\partial t} = -\nabla P_1 - \rho_1 g \hat{y}. \tag{6.7}$$

The line $y = 0$ is defined to be the unperturbed interface between air and water and the top of the glass is at $y = h$. The boundary condition for the water at the top would be

$$v_y = 0 \quad at \quad y = h. \tag{6.8}$$

Assume perturbations of the form

$$\mathbf{v}_1 = \mathbf{v}_1(y) e^{n + i\mathbf{k} \cdot \mathbf{x}}, \tag{6.9}$$

where $\mathbf{k}$ lies in the xz-plane and the positive γ implies instability. The incompressibility condition can be written as

$$\frac{\partial v_{1y}}{\partial y} + i\mathbf{k} \cdot \mathbf{v}_{1\perp} = 0, \tag{6.10}$$

where $\perp$ means perpendicular to the y-direction. The y- and $\perp$-components of Eq. (6.7) become, respectively,

$$\gamma \rho_0 v_{1y} = -\frac{\partial P_1}{\partial y} - \rho_1 g \tag{6.11}$$

$$\gamma \rho_0 \mathbf{v}_{1\perp} = -i\mathbf{k}p_1. \tag{6.12}$$

We'll take the dot product of Eq. (6.12) with $i\mathbf{k}$ and use Eq. (6.10) to eliminate $\mathbf{k} \cdot \mathbf{v}_{1\perp}$ to obtain

$$\gamma \rho_0 \frac{\partial v_{1y}}{\partial y} = k^2 P_1. \tag{6.13}$$

The perturbed density, as given in Eq. (6.6), is

$$\gamma \rho_1 = -v_{1y} \frac{\partial \rho_0}{\partial y}. \tag{6.14}$$

Now, ρ_1 and P_1 are substituted in Eq. (6.10) to obtain the eigenvalue problem

$$\frac{\partial}{\partial y} \left[\gamma^2 \rho_0 \frac{\partial v_{1y}}{\partial y} \right] = \left[\gamma^2 \rho_0 - g \frac{\partial \rho_0}{\partial y} \right] k^2 v_{1y}. \tag{6.15}$$

The above equation is solved for the interior and interface separately. For the interior, $\partial \rho_0 / \partial y = 0$ and $\rho_0 = $ constant, which means that Eq. (6.15) reduces to

$$\frac{\partial^2 v_{1y}}{\partial y^2} = k^2 v_{1y}, \tag{6.16}$$

with the solution satisfying the boundary condition as

$$v_{1y} = A \sinh[k(y-h)]. \tag{6.17}$$

For the interface, Eq. (6.15) should be integrated from $y = 0_-$ to $y = 0_+$ to obtain

$$\left[\gamma^2 \rho_0 \frac{\partial v_{1y}}{\partial y} \right]_{0_-}^{0_+} = -\left[g\rho_0 k^2 v_{1y} \right]_{0_-}^{0_+} \tag{6.18}$$

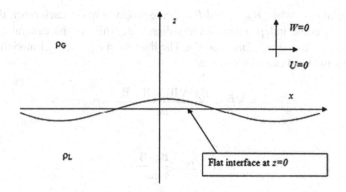

Fig. 6.1 A simple sketch of the Rayleigh–Taylor configuration

or

$$\gamma^2 \frac{\partial v_{1y}}{\partial y} = -gk^2 v_{1y}, \tag{6.19}$$

where all quantities refer to the upper (water) side of the interface, since by assumption $\rho_0(y = 0_-) \approx 0$. Substituting Eq. (6.17) into Eq. (6.19) leads to the dispersion relation

$$\gamma^2 = kg\tanh[k_\perp h]. \tag{6.20}$$

The above expression implies that the configuration is always unstable since $\gamma^2 > 0$. Also, one can realize that the most unstable ones are the short wavelengths. Taking into account other effects, such as surface tension, may stabilize the system for a given range of wavelengths. A simple model for the Rayleigh–Taylor instability is sketched in Fig. 6.1.

Let's now turn our attention to RT instability in magnetohydrodynamic fluids. We shall replace water by a magneto fluid (a fluid that satisfies the MHD equations) and atmospheric pressure by a vertical magnetic field whose gradient balances the gravitational force; that is, at each y, the upward force of $-\nabla b^2/2\mu_0$ supports the downward force of the weight of the plasma. In order for $-\nabla b^2$ to point upward in the y-direction, the magnetic field must depend on y such that its magnitude decreases with increasing y. Also, it is required that $B_y = 0$, so that ∇B^2 is perpendicular to the magnetic field and the field can be considered locally straight. The equilibrium magnetic field may be assumed to be

$$\mathbf{B}_0 = B_{x0}(y)\hat{x} + B_{z0}(y)\hat{z}. \tag{6.21}$$

The unit vector related to the equilibrium may be written as

$$\hat{B}_0 = \frac{B_{x0}(y)\hat{x} + B_{z0}(y)\hat{z}}{\sqrt{B_{x0}(y)^2 + B_{z0}(y)^2}}. \tag{6.22}$$

In the special case when $B_{x0}(y)$ and $B_{z0}(y)$ are proportional to each other, the field line would become independent of the y-direction, while in the general case, $\hat{B}_0$ depends on y, rotating as a function of y. The linearized equations of motion for the incompressible fluid can be written as

$$\rho_0 = -\nabla \bar{P}_1 + \frac{\mathbf{B}_0 \cdot \nabla \mathbf{B}_1 + \mathbf{B}_1 \cdot \mathbf{B}_0}{\mu_0} - \rho_1 g \hat{y}, \tag{6.23}$$

where

$$\bar{P}_1 = P_1 + \frac{\mathbf{B}_0 \cdot \mathbf{B}_1}{\mu_0}$$

is the perturbation of the total (hydrodynamic and magnetic) pressure, namely, $P + B^2/2\mu_0$. The components of Eq. (6.23) are given by

$$\gamma \rho_0 v_{1y} = -\frac{\partial \bar{P}}{\partial y} + \frac{i(\mathbf{k} \cdot \mathbf{B}_0) B_{1y}}{\mu_0} - \rho_1 g \tag{6.24}$$

$$\gamma \rho_0 \mathbf{v}_{1\perp} = -i\mathbf{k}\bar{P} + \frac{1}{\mu_0}\left[i(\mathbf{k} \cdot \mathbf{B}_0)\mathbf{B}_{1\perp} + B_{1y}\frac{\partial \mathbf{B}_0}{\partial y}\right]. \tag{6.25}$$

Taking the dot product of Eq. (6.25) with $i\mathbf{k}$ and using (6.10), we obtain

$$-\gamma\rho_0\frac{\partial v_{1y}}{\partial y} = k^2\bar{P} + \frac{1}{\mu_0}\left[-(\mathbf{k} \cdot \mathbf{B}_0)\mathbf{k} \cdot \mathbf{B}_{1\perp} + iB_{1y}\frac{\partial(\mathbf{k} \cdot \mathbf{B}_0)}{\partial y}\right]. \tag{6.26}$$

Assuming that the magnetic field is divergent-free, namely, $\nabla \cdot \mathbf{B}_1 = 0$, the perturbed perpendicular field can be written as

$$i\mathbf{k} \cdot \mathbf{B}_{1\perp} = -\frac{\partial B_{1y}}{\partial y}, \tag{6.27}$$

and so Eq. (6.26) reduces to

$$k^2\bar{P} = -\gamma\rho_0\frac{\partial v_{1y}}{\partial y} - \frac{1}{\mu_0}\left[-i(\mathbf{k} \cdot \mathbf{B}_0)\frac{\partial B_{1y}}{\partial y} + iB_{1y}\frac{\partial(\mathbf{k} \cdot \mathbf{B}_0)}{\partial y}\right]. \tag{6.28}$$

Eliminating $\bar{P}$ from Eq. (6.24) and substituting in the above equation leads to the following:

$$\gamma\rho_0 v_{1y} = -\frac{1}{k^2}\frac{\partial}{\partial y}\left[-\gamma\rho_0\frac{\partial v_{1y}}{\partial y} - \frac{1}{\mu_0}\left[-(i\mathbf{k} \cdot \mathbf{B}_0)\frac{\partial B_{1y}}{\partial y} + B_{1y}\frac{\partial(i\mathbf{k} \cdot \mathbf{B}_0)}{\partial y}\right]\right]$$

$$+ \frac{i(\mathbf{k} \cdot \mathbf{B}_0)B_{1y}}{\mu_0} - \rho_1 g. \tag{6.29}$$

We further make use of Ohm's law to obtain

$$\mathbf{E}_1 + \mathbf{v} \times \mathbf{B}_0 = 0. \tag{6.30}$$

Taking the curl and using Faraday's law, we obtain

$$\gamma \mathbf{B}_1 = \nabla \times [\mathbf{v}_1 \times \mathbf{B}_0]. \tag{6.31}$$

We take the dot product with $\hat{y}$ and use the vector identity $\nabla \cdot (\mathbf{F} \times \mathbf{G}) = \mathbf{G} \cdot \nabla \times \mathbf{F} - \mathbf{F} \cdot \nabla \times \mathbf{G}$ to obtain

$$\gamma B_{1y} = \hat{y} \cdot \nabla \times [\mathbf{v} \times \mathbf{B}_0] = \nabla \cdot [(\mathbf{v} \times \mathbf{B}_0) \times \hat{y}] = \nabla \cdot [v_{1y}\mathbf{B}_0] = i\mathbf{k} \cdot \mathbf{B}_0 v_{1y}. \tag{6.32}$$

Substituting into Eq. (6.29), using Eqs. (6.32) and (6.14), and rearranging the terms give

$$\frac{\partial}{\partial y}\left[\left[\gamma^2 \rho_0 + \frac{1}{\mu_0}(\mathbf{k} \cdot \mathbf{B}_0)^2\right]\frac{\partial v_{1y}}{\partial y}\right] = k^2\left[\gamma^2 \rho_0 - g\frac{\partial \rho_0}{\partial y} + \frac{(\mathbf{k} \cdot \mathbf{B}_0)^2}{\mu_0}v_{1y}\right]. \tag{6.33}$$

When $\mathbf{k} \cdot \mathbf{B}_0 = 0$, the above equation reduces to the hydrodynamic case. Integrating the equation from $y = 0$ to $y = h$, we obtain

$$\left[\left(\gamma^2 \rho_0 + \frac{1}{\mu_0}(\mathbf{k} \cdot \mathbf{B}_0)^2\right)v_{1y}\frac{\partial v_{1y}}{\partial y}\right]_0^h - \int_0^h\left[\gamma^2 \rho_0 + \frac{1}{\mu_0}(\mathbf{k} \cdot \mathbf{B}_0)^2\right]\left(\frac{\partial v_{1y}}{\partial y}\right)^2 dy$$

$$= k^2\int_0^h\left[\gamma^2 \rho_0 - g\frac{\partial \rho_0}{\partial y} + \frac{(\mathbf{k} \cdot \mathbf{B}_0)^2}{\mu_0}\right]v_{1y}^2 dy. \tag{6.34}$$

The first term on the left-hand side of the above equation vanishes on applying the boundary conditions, so that the value for γ^2 turns out to be

$$\gamma^2 = \frac{\int_0^h dy\left[k^2 g\frac{\partial \rho_0}{\partial y}v_{1y}^2 - \frac{(\mathbf{k} \cdot \mathbf{B}_0)^2}{\mu_0}(k^2 v_{1y}^2 + (\frac{\partial v_{1y}}{\partial y})^2)\right]}{\int_0^h dy\rho_0\left[k^2 v_{1y}^2 + (\frac{\partial v_{1y}}{\partial y})^2\right]}. \tag{6.35}$$

When $\mathbf{k} \cdot \mathbf{B}_0 = 0$ and the density gradient is positive everywhere, $\gamma^2 > 0$, and so this leads to instability. If the density gradient is negative everywhere except at a region of finite thickness $\triangle y$, then the system will be unstable with respect to an interchange at one of the strata. The velocity will be concentrated at this unstable stratum and so the integrands will vanish everywhere except at the unstable stratum, giving a growth rate of $\gamma^2 \sim g\triangle y\rho_0^{-1}\partial \rho_0/\partial y$, where $\partial \rho_0/\partial y$ is the value in the unstable region.

The hydromagnetic Rayleigh–Taylor instability is also referred to as the Kruskal–Schwarzschild instability. For nonzero $\mathbf{k} \cdot \mathbf{B}_0$, it opposes the effect of the destabilizing positive density gradient, reducing the growth rate to $\gamma^2 \sim g \triangle y \rho_0^{-1} \partial \rho_0 / \partial y - (\mathbf{k} \cdot \mathbf{B}_0)^2 / \mu_0$. Thus, a sufficiently strong field will stabilize the system.

Application: The launch of the Hinode satellite led to the discovery of rising plumes, dark in chromospheric lines, that propagate from large (10- mm) bubbles that form at the base of quiescent prominences. The plumes move through a height of approximately 10 mm while developing highly turbulent profiles. The magnetic Rayleigh–Taylor instability was hypothesized to be the mechanism that drives these flows. Recently, Hillier et al. (2012), using three-dimensional (3D) MHD simulations, investigated the nonlinear stability of the Kippenhahn–Schluter prominence model for the interchange mode of the magnetic Rayleigh–Taylor instability. Their model deals with the rise of a buoyant tube inside the quiescent prominence model, where the interchange of magnetic field lines becomes possible at the boundary between the buoyant tube and the prominence. The nonlinear interaction between plumes plays an important role for determining the plume dynamics. Using the results of ideal MHD simulations, they determined the initial parameters for the model and buoyant tube affect the evolution of instability. They found that the 3D mode of the magnetic Rayleigh–Taylor instability grows, creating upflows aligned with the magnetic field of constant velocity (maximum found at $7.3 \, \mathrm{km \, s^{-1}}$). The width of the upflows is dependent on the initial conditions, with a range of 0.5–4 mm, which propagate through heights of 3–6 mm. Another application of RT stability has been studied by Ali et al. (2009) in dense magnetoplasmas.

6.3 Kelvin–Helmholtz Instability

Low-frequency MHD waves that are excited at boundaries by velocity shears are called Kelvin–Helmholtz waves. Unlike the Rayleigh–Taylor instability, which occurs due to density discontinuities, the Kelvin–Helmholtz (KH) instabilities have to do with velocity shears. A tangential discontinuity supports velocity shears, and waves observed at such interfaces are the Kelvin–Helmholtz waves. These waves are also low-frequency surface waves that can grow and become unstable under certain conditions. This section studies such conditions for the KH instability.

Consider a boundary across which there is a sheared flow, and assume for simplicity that the two fluids are ideal (infinite conductivity, $\sigma = \infty$) and incompressible, and the pressure is isotropic. The equilibrium and perturbed quantities are denoted by

$$\mathbf{v} = \mathbf{V}_0 + \delta \mathbf{V}$$

$$p = p_0 + \delta p$$

$$\mathbf{B} = \mathbf{B}_0 + \delta \mathbf{B}. \tag{6.36}$$

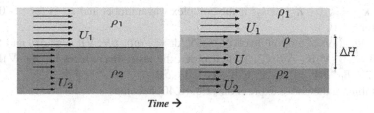

Fig. 6.2 A simple sketch for the Kelvin–Helmholtz instability; from Gramer (2007)

In a Cartesian coordinate system, let the z-axis be directed along the normal to the plane of the discontinuity. For a tangential discontinuity, $\mathbf{V}_0$ and $\mathbf{B}_0$ lie in the xy-plane. Assume that the perturbation produces waves propagating in the xy-plane with the condition that the waves may decay in strength away from the xy-plane, in the z-direction. A simple sketch that describes the Kelvin–Helmholtz instability configuration is presented in Fig. 6.2. Assuming perturbations of the form

$$\exp[i(k_x x + k_y y - \Omega t) - k_z z], \tag{6.37}$$

where Ω is the Doppler-shifted frequency of the wave as measured by a stationary observer in the frame of the boundary, the decay length is given by the reciprocal of k_z. Thus, for $z < 0$, $k_z < 0$, and vice versa. The MHD equations (ideal) are

$$\frac{\partial \mathbf{B}}{\partial t} = -\nabla \times \mathbf{E} \tag{6.38}$$

$$\mathbf{E} = -\mathbf{V} \times \mathbf{B} \tag{6.39}$$

$$\rho_m \frac{d\mathbf{V}}{dt} = -\nabla p + \mathbf{J} \times \mathbf{B}. \tag{6.40}$$

For simplicity, we assume the pressure to be a scalar. The magnetic field and the current density are related to each other by the relation $\nabla \times \mathbf{B} = \mu_0 \mathbf{J}$. Assuming plane wave solution to Eq. (6.38) results in

$$
\begin{aligned}
-i\Omega \delta \mathbf{B} &= (\mathbf{B} \cdot \nabla)\mathbf{V} - (\mathbf{V} \cdot \nabla)\mathbf{B} \\
&= (\mathbf{B}_0 \cdot \nabla)\delta \mathbf{V} - (\mathbf{V}_0 \cdot \nabla)\delta \mathbf{B} \\
&= i\mathbf{k}_t \cdot (B_0 \delta \mathbf{V} - V_0 \delta \mathbf{B}).
\end{aligned} \tag{6.41}
$$

The Doppler-shifted frequency is $\Omega = \omega + \mathbf{k} \cdot \mathbf{V}_0$, so that the above equation can be written as

$$\omega \delta \mathbf{B} = -(\mathbf{B}_0 \cdot \mathbf{k}_t) \delta \mathbf{V}. \tag{6.42}$$

The divergence-free (incompressible fluid) condition $\nabla \cdot \mathbf{V} = 0$ yields

$$\kappa_\pm \cdot \delta \mathbf{V} = 0, \tag{6.43}$$

where $\kappa_{\pm} = (k_x\hat{x} + K_y\hat{y}) \pm k_z\hat{z}$ is the complex wavenumber and κ_+ is for $z > 0$ and κ_- is for $z < 0$.

The equation of motion involves $\rho_m dV/dt = -\omega\rho_m\delta V$, $\nabla p = \kappa_{\pm}\delta p$, and $(\nabla \times B) \times B = (B \cdot \nabla)B - \nabla B^2/2$. To first order, the last term reduces to $(B_0 \cdot \nabla)B_1 + \nabla(B_0^2/2 + B_0 \cdot \delta B) = (B_0 \cdot \kappa_{\pm})\delta B - i\kappa_{\pm}(B_0 \cdot \delta B)$.

Combining the above expressions, Eq. (6.40) becomes

$$\omega\rho_m\delta V - \kappa_{\pm}\delta p = -\frac{(B_0 \cdot k_t)\delta B}{\mu_0} + \frac{(B_0 \cdot \delta B)\kappa_{\pm}}{\mu_0}. \tag{6.44}$$

The total pressure is $p^{\star} = p + B2/2\mu_0$, and to first order

$$\delta p^{\star} = \delta p + \frac{B_0 \cdot \delta B}{\mu_0}. \tag{6.45}$$

Incorporating Eqs. (6.44) and (6.42) results in

$$\kappa_{\pm}\delta p^{\star} = \frac{\rho_m}{\omega}\left[\omega^2 - \frac{(B_0 \cdot k_t)^2}{\mu_0\rho_m}\right]\delta V. \tag{6.46}$$

Taking the scalar product of the above equation with $\kappa_{\pm}$ and noting that $\kappa_{\pm} \cdot \delta V = 0$, we obtain

$$\kappa_{\pm}^2\delta p^{\star} = 0. \tag{6.47}$$

The solution corresponding to

$$\delta p^{\star} = 0$$

yields the intermediate Alfvén mode, namely, $\omega/k = \pm V_A$, where the Alfvén speed $V_A = B_0\lambda/\sqrt{\mu_0\rho_m}$ or

$$\kappa_{\pm}^2 = 0.$$

The second option is related to surface waves, yielding

$$k_t^2 = k_z^2. \tag{6.48}$$

Equation (6.48) is a condition for the decay length of the perturbation away from the surface of discontinuity; that is, $k_z^{-1} = k_t^{-1}$. Scalar multiplication of Eq. (6.46) with $\hat{z}$ yields

$$\delta p_t\kappa_{\pm} \cdot \hat{z} = \rho_m\left[\omega^2 - \frac{(B_0 \cdot k_t)^2}{\mu_0\rho_m}\right]\delta V \cdot \hat{z}. \tag{6.49}$$

It is important to note that

$$\kappa_{\pm} \cdot \hat{z} = \pm ik_z \tag{6.50}$$

and

$$\delta \mathbf{V} \cdot \hat{\mathbf{z}} = \frac{d\delta z}{dt}$$

$$= \frac{\partial \delta z}{\partial t} + (\mathbf{V}_0 \cdot \nabla)\delta z$$

$$= i(\Omega - k\mathbf{V}_0)\delta z$$

$$= i\omega \delta z. \tag{6.51}$$

Thus, Eq. (6.49) becomes

$$\pm k_z \delta p^\star = [\omega^2 - (\mathbf{V}_A \cdot \mathbf{k})^2]\rho_m \delta z. \tag{6.52}$$

The total pressure across a tangential discontinuity is continuous; namely, $[p^\star] = 0$. Also, the displacement of the two fluids in the z-direction must be continuous in order to avoid separation or interpenetration of the fluids: $[\delta z] = 0$. Equation (6.52) has the form $\delta p^\star = A\delta z$. Thus, $[\delta p^\star] = [A\delta z] = [A][\delta z[= 0$, which implies that $[A] = 0$, which, written explicitly, will look like

$$\rho_1 \left[\omega_1^2 - (\mathbf{V}_{A1} \cdot \mathbf{k})^2\right] + \rho_2 \left[\omega_2^2 - (\mathbf{V}_{A2} \cdot \mathbf{k})^2\right] = 0. \tag{6.53}$$

The frequency measured by an observer is $\Omega = \omega_1 + \mathbf{V}_1 \cdot \mathbf{k} = \omega_2 + \mathbf{V}_2 \cdot \mathbf{k}$ and is the same on both sides. Solving for Ω yields

$$\Omega = \frac{\rho_1 \mathbf{V}_1 \cdot \mathbf{k} + \rho_2 \mathbf{V}_2 \cdot \mathbf{k}}{(\rho_1 + \rho_2)}$$

$$\pm \frac{1}{(\rho_1 + \rho_2)} \left[(\rho_1 + \rho_2)\left[\rho_1(\mathbf{V}_{A1} \cdot \mathbf{k})^2 + \rho_2(\mathbf{V}_{A2} \cdot \mathbf{k})^2\right] - \rho_1\rho_2(\triangle \mathbf{V} \cdot \mathbf{k})^2\right]^{1/2},$$

$$\tag{6.54}$$

where $\triangle \mathbf{V} = \mathbf{V}_1 - \mathbf{V}_2$ is the basic shear and the subscripts 0 and m have been omitted from the velocity and density.

The relation given by Eq. (6.54) is the dispersion relation for the Kelvin–Helmholtz waves. These waves are unstable when Ω is imaginary. Equation (6.54) shows that when

$$\rho_1\rho_2(\triangle \mathbf{V} \cdot \mathbf{k})^2 > \frac{1}{\mu_0}(\rho_1 + \rho_2)[(\mathbf{B}_1 \cdot \mathbf{k})^2 + (\mathbf{B}_2 \cdot \mathbf{k})^2], \tag{6.55}$$

the imaginary part of Eq. (6.54), $\mathrm{Im}(\Omega) > 0$.

In order to have instability, a shear threshold is essential. This threshold is required because the tension in the magnetic field will resist any force acting on it to stretch it. Thus, the growth of the wave will occur only if the velocity shear overcomes the magnetic tension. If $\mathbf{B}_1$ and $\mathbf{B}_2$ are perpendicular to $\mathbf{k}$, then the right-hand side of Eq. (6.55) vanishes and $(\triangle \mathbf{V} \cdot \mathbf{k})^2 > 0$ implies that the boundary is unstable to an arbitrary small shear across the boundary. It is interesting to note that the growth rate depends on the relative directions of $\mathbf{B}$, $\mathbf{k}$, and $\triangle \mathbf{V}$. If α is the angle between $\triangle \mathbf{V}$ and $\mathbf{k}$, and θ_1 and θ_2 are the angles between $\mathbf{k}$, $\mathbf{B}_1$, and $\mathbf{B}_2$, then the instability criterion can be written as

$$\triangle U^2 \cos^2 \alpha > \frac{1}{\mu_0} \left(\frac{1}{\rho_1} + \frac{1}{\rho_2} \right) \left(B_1^2 \cos^2 \theta_1 + B_2^2 \cos^2 \theta_2 \right). \tag{6.56}$$

Application: One of the important applications of the KH instability is in coronal streamers and may also be found in solar wind, where the plasma streams relative to the surrounding plasma, which is relatively stationary. Flows and instabilities play a major role in the dynamics of magnetized plasmas, including the solar corona, magnetospheric and heliospheric boundaries, cometary tails, and astrophysical jets. The nonlinear effects, multiscale, and microphysical interactions inherent to the flow-driven instabilities are believed to play a role, for example, in plasma entry across a discontinuity, generation of turbulence, and enhanced drag. However, in order to clarify the efficiency of macroscopic instabilities in these processes, we lack proper knowledge of their overall morphological features. Foullon et al. (2011) reported the first observations of the temporally and spatially resolved evolution of the magnetic Kelvin–Helmholtz instability in the solar corona. Unprecedented high-resolution imaging observations of vortices developing at the surface of a fast coronal mass ejecta taken from the new Solar Dynamics Observatory validate theories of the nonlinear dynamics involved. The new findings are a cornerstone for developing a unifying theory on flow-driven instabilities in rarefied magnetized plasmas, which is important for understanding the fundamental processes in key regions of the Sun–Earth system. Ofman and Thompson (2011) reported Solar Dynamics Observatory (SDO) observation of Kelvin–Helmholtz instability in the solar corona.

6.4 Parametric Instability

Parametric instability is one of the most thoroughly investigated nonlinear wave–wave interactions. The theory is basically linear, but linear about an oscillating equilibrium. A standard explanation for the parametric instability is that of two coupled oscillators M_1 and M_2 on a bar resting on a pivot. The pivot has the freedom to move back and forth with a frequency ω_0, whereas the natural frequencies of the oscillators are ω_1 and ω_2, respectively. In the absence of friction, the pivot does not encounter any resistance as long as the masses M_1 and M_2 are not in motion. If P does not move, while M_2 is set to motion, it will induce movement to M_1, as long as the natural frequency of M_1 is not ω_2, and the amplitude will be small. If both P and M_2 are set in motion, then the displacement of M_1 as a function of time will be

$$\cos(\omega_2 t)\cos(\omega_0 t) = (1/2)\cos[(\omega_2 + \omega_0)t] + (1/2)\cos[(\omega_2 - \omega_0)t]. \tag{6.57}$$

In the event of ω_1 equaling either $\omega_2 + \omega_0$ or $\omega_2 - \omega_0$, M_1 will get excited resonantly, which will result in the growth of the amplitude. With M_1 oscillating, M_2 will gain energy because one of the beat frequencies of ω_1 with ω_0 is nothing but ω_2. Thus, if one of the oscillators starts moving, then the other will get excited, which will result in the system becoming unstable. Depending on the energy coming from P, the oscillation amplitude is unaffected by M_1 and M_2, and the instability can be treated in the linear regime. In a plasma or MHD, P, M_1, M_2 may be different types of waves.

Let's work out a more quantitative analysis pertaining to parametric instabilities by considering the equations of motion of two harmonic oscillators x_1 and x_2 as follows:

$$\frac{d^2 x_1}{dt^2} + \omega_1^2 x_1 = 0, \tag{6.58}$$

where ω_1 is its resonant frequency. If it is driven by a time-dependent force that is proportional to the product of the amplitude E_0 of the driver or pump, and the amplitude x_2 of the second oscillator, the equation of motion becomes

$$\frac{d^2 x_1}{dt^2} + \omega_1^2 x_1 = c_1 x_2 E_0, \tag{6.59}$$

where c_1 is a constant that indicates the strength of the mode coupling. One can write a similar equation for x_2 as follows:

$$\frac{d^2 x_2}{dt^2} + \omega_2^2 x_2 = c_2 x_1 E_0. \tag{6.60}$$

Assume that $x_1 = \bar{x}_1 \cos(\omega t)$, $x_2 = \bar{x}_2 \cos(\omega' t)$, and $E_0 = \bar{E}_0 \cos(\omega_0 t)$. Equation (6.60) reduces to

$$(\omega_2^2 - \omega'^2)\bar{x}_2 \cos(\omega' t) = c_2 \bar{E}_0 \bar{x}_1 \cos(\omega_0 t)\cos(\omega t)$$
$$= c_2 \bar{E}_0 \bar{x}_1 (1/2)\left[\cos(\omega_0 + \omega)t] + \cos[(\omega_0 - \omega)t]\right]. \tag{6.61}$$

The driving terms on the right-hand side can excite oscillators x_2 with frequencies

$$\omega' = \omega_0 \pm \omega. \tag{6.62}$$

In the absence of nonlinear interactions, x_2 can have only the frequency ω_2, so that we have $\omega' = \omega_2$. However, the driving terms can cause a shift in the frequency so that $\omega' \approx \omega_2$. Also, ω' can be complex, because of the damping, or it can grow, leading to instability. In both cases, x_2 is an oscillator with a finite amplitude and can respond to a range of frequencies about ω_2. If ω is small, then it is evident from Eq. (6.62) that both choices for ω' may lie within the bandwidth of x_2, and one must in principle give allowance for two oscillators, $x_2(\omega_0 + \omega)$ and $x_2(\omega_0 - \omega)$. Inserting the new variation for x_1 and x_2 as $x_1 = \bar{x}_1 \cos(\omega^* t)$ and $x_2 = \bar{x}_2 \cos[(\omega_0 \pm \omega)t]$ into Eq. (6.59), we have

$$(\omega_1^2 - \omega^{*2})\bar{x}_1 \cos(\omega^* t) = c_1 \bar{E}_0 \bar{x}_2 (1/2)(\cos[(\omega_0 + (\omega_0 \pm \omega)t]$$

$$+ \cos[(\omega_0 - (\omega_0 \pm \omega))]t)$$

$$= c_1 \bar{E}_0 \bar{x}_2 (1/2)[\cos(2\omega_0 \pm \omega)t + \cos\omega t]. \quad (6.63)$$

The driving terms excite not only the original oscillation $x_1(\omega)$ but also the frequencies $\omega^* = 2\omega_0 \pm \omega$. Consider the case when $|\omega_0| \gg |\omega_1|$, so that $2\omega_0 \pm \omega$ lies outside the range of frequencies to which x_1 would respond, and neglect $x_1(2\omega_0 - \omega)$. Thus, we have three oscillators: $x_1(\omega)$, $x_2(\omega_0 - \omega)$, and $x_2(\omega_0 + \omega)$, which are coupled.

The dispersion relation is obtained by setting the determinant of the coefficients equal to zero as follows:

$$\begin{pmatrix} \omega^2 - \omega_1^2 & c_1 E_0 & c_1 E_0 \\ c_2 E_0 & (\omega_0 - \omega)^2 - \omega_2^2 & 0 \\ c_2 E_0 & 0 & (\omega_0 + \omega)^2 - \omega_2^2 \end{pmatrix} = 0. \quad (6.64)$$

A solution of the above dispersion relation with $\text{Im}(\omega) > 0$ will lead to instability.

For small frequency shifts and small damping or growth rates, we can set ω and ω' approximately equal to the undisturbed natural frequencies ω_1 and ω_2. Equation (6.62) give a frequency-matching condition as

$$\omega_0 \approx \omega_2 \pm \omega_1. \quad (6.65)$$

If we interpret the oscillators as waves in a plasma, then we must replace ωt by $\omega t - \mathbf{k} \cdot \mathbf{r}$. The corresponding wavelength-matching condition would become

$$\mathbf{k}_0 = \mathbf{k}_2 \pm \mathbf{k}_1, \quad (6.66)$$

which will describe spatial beats. This will imply the periodicity of points of constructive and destructive interference in space. Parametric instabilities will occur at any amplitude if damping is not present. However, in practice, a small amount of either the collisional or Landau damping will prevent instability, unless the pump of the wave is very strong. In such a situation, one can introduce damping Γ_1 and Γ_2 for the oscillators x_1 and x_2, with a change in the equation as

$$\frac{d^2 x_1}{dt^2} + \omega_1^2 x_1 + 2\Gamma_1 \frac{dx_1}{dt} = 0. \quad (6.67)$$

In this case, the threshold for stability will be

$$c_1 c_2 (E_0^2)_{\text{thresh}} = 4\omega_1 \omega_2 \Gamma_1 \Gamma_2, \quad (6.68)$$

and the threshold goes to zero with the damping of either of the waves.

Application: Low-frequency turbulence in the solar wind is characterized by a high degree of Alfvenicity close to the Sun. Cross-helicity, which is a measure of Alfvenic correlation, tends to decrease with increasing distance from the Sun at high latitudes as well as in slow-speed streams at low latitudes. However, large-scale inhomogeneities (velocity shears, the heliospheric current sheet) are present, which are sources of decorrelation; moreover, at high latitudes, the wind is much more homogeneous, and a possible evolution mechanism may be represented by the parametric instability. The parametric decay of a circularly polarized broadband Alfven wave has been investigated by Malara et al. (2001). The time evolution has been obtained by numerically integrating the full set of nonlinear MHD equations, up to instability saturation. They find that, for $\beta \approx 1$, the final cross-helicity is ≈ 0.5, corresponding to a partial depletion of the initial correlation. Compressive fluctuations at a moderate level are also present. Most of the spectrum is dominated by forward-propagating Alfvenic fluctuations, while backscattered fluctuations dominate large scales.

The parametric decay of circularly polarized Alfven waves with multidimensional simulations in periodic and open domains has been studied by Del Zanna et al. (2001). For higher values of beta, they found that the cross-helicity decreases monotonically with time toward zero, implying an asymptotic balance between inward and outward Alfvenic modes, a feature similar to the observed decrease with distance in the solar wind. Although the instability mainly takes place along the propagation direction, in the two- and three-dimensional cases, a turbulent cascade occurs also transverse to the field. The asymptotic state of density fluctuations appears to be rather isotropic, whereas a slight preferential cascade in the transverse direction is seen in magnetic field spectra. Finally, parametric decay is shown to occur also in a nonperiodic domain with open boundaries, when the mother wave is continuously injected from one side. In two and three dimensions, a strong transverse filamentation is found at long times, reminiscent of density ray-like features observed in the extended solar corona and pressure-balanced structures found in solar wind data.

The parametric interaction of beam-driven Langmuir waves in the solar wind was studied by Gurnett et al. (1981), while Weatherall et al. (1981) studied parametric instabilities in weakly magnetized plasmas.

6.5 Parker Instability

The concept of buoyancy stability is similar to convective instability and comes about due to the magnetic field providing the pressure without any mass. Thus, in pressure equilibrium, matter containing magnetic field is lighter than matter without, and it operates even when the fluid is incompressible. In most of the astronomical situations, fluids are not incompressible. Thus, the modes by which buoyancy can drive an instability are drastically different from that of the incompressible case. This is the case of instability discussed by Parker (1979), and we shall briefly describe it here (also see Pringle and King 2007).

Consider a stratified fluid with gravity, given by $\mathbf{g} = (0,0,-g)$, and a horizontal magnetic field, $\mathbf{B}_0 = (0, B_0(z), 0)$, in the y-direction. This implies that $\nabla \times \mathbf{B}_0 \neq 0$ although $\nabla \cdot \mathbf{B}_0 = 0$, and a current proportional to $-\partial B_0/\partial z$ in the x-direction is present. The equation of state of the unperturbed fluid is assumed to be isothermal, so that the relation between the density and pressure is given by $p = c_S^2 \rho$, where c_S is the isothermal sound speed. Another simplification is that the magnetic field is such that the magnetic pressure is a constant fraction α of the gas pressure. This means that $B_2^2/2 = \alpha p$. The hydrostatic equilibrium equation is given by

$$\frac{d}{dz}\left(p + \frac{1}{2}B^2\right) = -\rho g. \tag{6.69}$$

The solution of the above equation is as follows:

$$p(z) = p_0 \exp(-z/H) \tag{6.70}$$

$$\rho(z) = \rho_0 \exp(-z/H) \tag{6.71}$$

and

$$B_0(z) = B_{00}\exp(-z/2H), \tag{6.72}$$

where the scale height H is given by

$$H = \frac{(1+\alpha)c_S^2}{g}. \tag{6.73}$$

Here p_0, ρ_0, and B_{00} are all constants, at the level $z = 0$.

The linearized perturbations are given by (to begin with the mass conservation)

$$\frac{\partial \rho'}{\partial t} + w\frac{d\rho}{dz} = -\rho \nabla \cdot \mathbf{v}. \tag{6.74}$$

Assuming the perturbations to be adiabatic, that is, $\delta p = (\gamma p/\rho)\delta\rho$, the mass conservation equation can be simplified to yield

$$\frac{\partial p'}{\partial t} = \gamma c_S^2 \frac{\partial \rho'}{\partial t} - \frac{(\gamma-1)c_S^2 \rho w}{H}. \tag{6.75}$$

The linearized induction equation is

$$\frac{\partial \mathbf{B}}{\partial t} = \nabla \times (\mathbf{v} \times \mathbf{B}_0), \tag{6.76}$$

which has three components given by

$$\frac{\partial B_x}{\partial t} = B_0 \frac{\partial u}{\partial y} \tag{6.77}$$

$$\frac{\partial B_y}{\partial t} = -B_0 \frac{\partial u}{\partial x} - B_0 \frac{\partial w}{\partial z} + B_0 \frac{w}{H} \tag{6.78}$$

$$\frac{\partial B_z}{\partial t} = B_0 \frac{\partial w}{\partial y}. \tag{6.79}$$

The linearized momentum equation is given by

$$\rho \frac{\partial \mathbf{v}}{\partial t} = -\nabla p' - \mathbf{B}_0 \times (\nabla \times \mathbf{B}) - \mathbf{B} \times (\nabla \times \mathbf{B}_0) + \rho' g, \tag{6.80}$$

with the following components:

$$\rho \frac{\partial u}{\partial t} = -\frac{\partial p'}{\partial x} + B_0 \left[\frac{\partial B_x}{\partial y} \right] \tag{6.81}$$

$$\rho \frac{\partial v}{\partial t} = -\frac{\partial p'}{\partial y} - B_0 \left[\frac{B_z}{2H} \right] \tag{6.82}$$

$$\rho \frac{\partial w}{\partial t} = -\frac{\partial p'}{\partial z} + B_0 \left[\frac{\partial B_z}{\partial y} - \frac{\partial B_y}{\partial z} + \frac{B_y}{2H} \right] - g\rho'. \tag{6.83}$$

Taking the Fourier transform of all the variables as $\propto \exp(i\omega t + i\mathbf{k} \cdot \mathbf{r})$ leads to linear, homogeneous, algebraic equations whose determinant will have to be calculated to determine the dispersion relation. The novel idea suggested by Parker is as follows: Assume

$$p' \propto \exp(i\omega t + i\mathbf{k} \cdot \mathbf{r}) \times \exp(-z/2H) \tag{6.84}$$

$$\rho' \propto \exp(i\omega t + i\mathbf{k} \cdot \mathbf{r}) \times \exp(-z/2H) \tag{6.85}$$

$$\mathbf{v} \propto \exp(i\omega t + i\mathbf{k} \cdot \mathbf{r}) \times \exp(+z/2H) \tag{6.86}$$

and

$$\mathbf{B} \propto \exp(i\omega t + i\mathbf{k} \cdot \mathbf{r}). \tag{6.87}$$

The above assumption will ensure that both the magnetic energy perturbation B^2 and the kinetic energy perturbation ρv^2 are independent of z. Before deriving the dispersion relation, we'll define a dimensionless frequency $\Omega^\star$ in terms of the time for an isothermal wave to cross a scale height as

$$\Omega^\star = \frac{\omega H}{c_S}, \tag{6.88}$$

and a dimensionless wave vector, in terms of the scale height as

$$\mathbf{q} = H\mathbf{k}. \tag{6.89}$$

The full dispersion relation is given by

$$\Omega^{*4} - \Omega^{*2}(2\alpha+\gamma)\left[q_y^2+q_z^2+1/4\right] + q_y^2\left[2\alpha\gamma(q_y^2+q_z^2+1/4)-(1+\alpha)(1+\alpha-\gamma)\right]$$

$$+\frac{q_x^2}{2\alpha(q_x^2+q_y^2)-\Omega^{*2}}\left[\gamma\Omega^{*4}-\Omega^{*2}\left[2\alpha\gamma q_y^2-2\alpha(2\alpha+\gamma)q_z^2+(\gamma-1)+(1/2)\alpha\gamma\right]\right.$$

$$\left.-4\alpha^2\gamma q_y^2(q_z^2+1/4)\right]. \tag{6.90}$$

The dispersion relation is a sixth-order equation in Ω^* and by time symmetry a cubic in Ω^{*2}.

The discussion on the dispersion relation will be for the specific case of modes with $k_y = 0$ and $k_x = 0$ and $k_y \neq 0$ instead of the full discussion on the acoustic, buoyant, and Alfvén waves. First, let's discuss the $k_y = 0$ case. In this case, the unperturbed magnetic field lies in the y-direction. If we set $k_y = 0$, the dispersion relation now reduces to

$$\Omega^{*2}\left[\Omega^{*4} - \Omega^{*2}(2\alpha+\gamma)(q_x^2+q_z^2+1/4) + q_x^2(\alpha(\alpha+\gamma)+\gamma-1)\right] = 0. \tag{6.91}$$

Two of the modes pertaining to $\Omega^{*2} = 0$ have zero frequency, and so are neutrally stable, the reason being the perturbation does not bend or twist the magnetic field lines. This implies that there are no magnetic waves. The remaining equation has the form

$$\Omega^{*4} - B\Omega^{*2} + C = 0, \tag{6.92}$$

where $B > 0$, implying that the sum of the roots is positive. The instability is set up if and only if $C < 0$, which means

$$\gamma < 1 - \alpha. \tag{6.93}$$

For an isothermal unperturbed fluid, it is known that in the absence of the field ($\alpha = 0$), it is unstable to convection if and only if $\gamma < 1$, which does not usually occur in fluids. If the field is included, $\alpha > 0$, we will require a very small value of γ to have instability. This implies that the presence of a magnetic field tends to stabilize the fluid. The equation describing the dragging of field lines is written as

$$\frac{D}{Dt}\left(\frac{B}{\rho}\right) = 0. \tag{6.94}$$

As the fluid moves, the local field varies as $B \propto \rho$, which implies that the magnetic pressure p_M varies as ρ^2 and the magnetic field acts like a gas with $\gamma = 2$.

Let's look at the modes with $k_x = 0, k_y \neq 0$. The dispersion relation in this case reduces to

$$\Omega^{*4} - \Omega^{*2}(2\alpha + \gamma)(q_y^2 + q_z^2 + 1/4)$$

$$- q_y^2 \left[(1+\alpha)(1+\alpha-\gamma) - 2\alpha\gamma(q_y^2 + q_z^2 + 1/4) \right]. \quad (6.95)$$

The factor Ω^{*2}, which corresponds to the two fast magnetosonic modes, has been removed. The remaining expression is written as

$$\Omega^{*4} - B\Omega^{*2} + C = 0, \quad (6.96)$$

with $B > 0$. Again, for this case, the instability exists if and only if $C < 0$, which implies that

$$q_y^2 + q_z^2 < \frac{(1+\alpha)(1+\alpha-\gamma)}{2\alpha\gamma} - \frac{1}{4}. \quad (6.97)$$

It is easy to realize that there is a nonvanishing range of unstable modes with $q_y^2 + q_z^2 > 0$ if and only if

$$\gamma < \frac{(1+\alpha)^2}{1+3\alpha/2}. \quad (6.98)$$

The right-hand side of the above expression is a monotonically increasing function of α whenever $\alpha > 0$, so that for any value of γ, there is instability for some large enough value of α, namely, for a sufficiently large enough magnetic field. Instability occurs if the gravitational energy released is more than the magnetic energy to undulate the field lines.

Application: The Sun is a magnetic star whose cyclic activity is thought to be linked to internal dynamo mechanisms. A combination of numerical modeling with various levels of complexity is an efficient and accurate tool to investigate such intricate dynamical processes. Jouve et al. (2010) investigated the role of the magnetic buoyancy process in 2D Babcock–Leighton dynamo models, by modeling more accurately the surface source term for the poloidal field. They incorporate, in mean-field models, the results of full 3D MHD calculations of the nonlinear evolution of a rising flux tube in a convective shell. More specifically, the Babcock–Leighton source term has been modified to take into account the delay introduced by the rise time of the toroidal structures from the base of the convection zone to the solar surface. They found that the time delays introduced in the equations produce a large temporal modulation of the cycle amplitude even when strong and thus rapidly rising flux tubes are considered. Aperiodic modulations of the solar cycle appear after a sequence of period-doubling bifurcations typical of nonlinear systems. The strong effects introduced even by small delays were found to be due to the dependence of the delays on the magnetic field strength at the base of the convection zone, the modulation being much less when the time delays remain constant. The influence on the cycle period, except when the delays are made artificially strong, is less significant. The modulated activity and the resulting butterfly diagram are more in accordance with observations than with the standard Babcock–Leighton model predictions. Chang and Quataert (2010) studied buoyancy instabilities in degenerate, collisional, magnetized plasmas with applications to compact stars, such as white dwarfs and neutron stars.

One of the interesting phenomena observed on the Sun is the coronal mass ejection (CME). This is a large-scale phenomenon that occurs at least once a day during solar activity. Theoretical models of the CME are based on emerging flux, which are bipolar in nature, arising from the subphotospheric layers, rising above the chromosphere, into the corona. One of the mechanisms that explains the rising of the flux tube is due to the Parker instability (also referred to as the magnetic buoyancy instability). One of the explanations for the mass ejection is due to twisting of the magnetic fields inside the flux tube and also to a magnetic reconnection that takes place inside the tube.

Chapter 7
Waves in the Sun

In Chaps. 4 and 5, we discussed theoretical formulations for the waves and oscillations in homogeneous as well as nonhomogeneous media. In this chapter, we will show applications of the different modes, such as the Alfvén, magnetoacoustic, and gravity modes, as relevant to our nearest neighbor, the Sun. Chapter 1, an introduction to the Sun, dealt with the physical properties of the interior as well as the outer atmosphere of the Sun. In this chapter, to begin with, we shall deal with the application of waves to the outer atmosphere of the Sun. We shall concentrate on 5-min oscillations, oscillations in sunspots, and chromospheric and coronal oscillations. We shall also briefly discuss the importance of these oscillations in deriving certain physical plasma parameters of the corona, using the tool of magnetoseismology. The importance of EIT (Extreme ultraviolet Imaging Telescope) and Moreton waves will be briefly presented. The importance of global modes of the Sun, with the motivation of understanding the interior of the Sun, as a function of the internal depth will be discussed in Chap. 8.

7.1 Five-Minute Oscillations

The first definite observations of solar oscillations were made by Leighton et al. (1962), who detected roughly periodic oscillations in Doppler velocity with periods of about 5 min. Evans and Michard (1962) confirmed the initial observations. The early observations were of limited duration, and the oscillations were generally interpreted as phenomena in the solar atmosphere. Later observations that resulted in power spectra as a function of wavenumber (e.g., Frazier 1968) indicated that the oscillations may not be mere surface phenomena. The first major theoretical advance in the field came when Ulrich (1970) and Leibacher and Stein (1971) proposed that the oscillations were standing acoustic waves in the Sun and predicted that power should be concentrated along ridges in a wavenumber–frequency diagram. Wolf (1972) and Ando and Osaki (1975) strengthened the hypothesis of standing waves by showing that oscillations in the observed frequency and wavenumber range may

A. Satya Narayanan, *An Introduction to Waves and Oscillations in the Sun*, Astronomy and Astrophysics Library, DOI 10.1007/978-1-4614-4400-8_7,
© Springer Science+Business Media New York 2013

be linearly unstable and hence can be excited. Acceptance of this interpretation of the observations as normal modes of solar oscillations was the result of the observations of Deubner (1975), which first showed ridges in the wavenumber–frequency diagram. Rhodes et al. (1977) reported similar observations. These, however, did not resolve the individual modes of solar oscillations, despite that these data were used to draw initial inferences about solar structure and dynamics. Using Doppler-velocity observations integrated over the solar disk, Claverio et al. (1979) were able to resolve the individual modes of oscillations corresponding to the largest horizontal wavelength. They found a series of almost equidistant peaks in the power spectrum, just as was expected from theoretical models. However, helioseismology as we know it today did not begin until Duvall and Harvey (1983) determined frequencies of a reasonably large number of solar oscillation modes covering a wide range of horizontal wavelengths. Since then, many sets of solar oscillation frequencies have been published.

Several theories were put forward to explain the observed 5-min oscillations. In 1975, Deubner found clear evidence of the true nature of these modes, wherein the relationship between the frequency and the horizontal wavenumber formed distinctive patterns. A simple relationship between the wavenumber and the frequency of the form $\omega^2 \approx k_\perp$ with a clustering and the bulk power falling in a period band centered about 5 min. In the early works of Ulrich (1970) and Leibacher and Stein (1971), such parabolic profiles were discussed. Thus, the subject of helioseismology was born, and in recent years, many landmark improvements, both in theory and in observations, have taken place, with the result that seismic determination of sound speed in the solar interior is understood reasonably well. The other aspects of the solar interior, such as differential rotation, magnetic field strength, and sunspot structure, have great connection with the topic of helioseismology.

Among the theoretical models put forward to explain the 5-min oscillations, two principal models are noteworthy. These models, which were proposed to explain the observed oscillations, depend on the rapidly changing thermal structure of the surface layers. The first class is that of an oscillation that is "ringing" of the stable photosphere, constantly struck from below by the convective elements. The first preliminary work on this idea was reported as early as the last century by Lamb (1909) and further developed from an astrophysical context by Meyer and Schmidt (1967); Schmidt and Zirker (1963); Stix (1970), and Schmidt and Stix (1973). The impulsive excitation causes an upward-propagating pulse. However, calculations of Ulrich (1970) and Stein and Schwarz (1972) show that the numerical value of the cutoff frequency of the acoustic modes is 180–220 s in the photosphere and lower chromosphere, rather than the 300 s period. The observation of Deubner (1973) shows several instances of high-frequency (period $\approx$200 s) wave trains, initiated by the appearance of bright granules. The second class that tries to explain the 5-min oscillations involves the trapping of waves in a cavity, resulting in an enhanced response to broadband excitation at the infrequencies of the cavity. Within this class, there are models that try to explain the phenomenon of trapped waves. A resonant cavity is a layer in which waves can in principle propagate vertically, bounded above and below by nonpropagating layers to reflect the waves.

It has been suggested that the resonant absorption of the 5-min oscillations in the chromosphere of the Sun may be responsible for both the sharp temperature increase in the upper chromosphere and the discrepancies between the theoretically calculated and observed frequencies of the 5-min oscillations (Zhukov 1992). Subsequently, Zukhov (1997) extended the earlier results by taking into account the resonant absorption in the canopy region of the magnetic field for a more realistic model of the Sun. The basic assumptions of this model are as follows: (1) The magnetic field is ignored in the bottom layer of the two-layer model; (2) the temperature falls with height; and (3) the top layer has a uniform magnetic field, with a sharp increase in the temperature up to coronal heights. The conclusion is that taking into account the acoustic energy absorption at Alfvén levels in the chromosphere, one can provide an explanation for the discrepancies between the observed and theoretically calculated frequencies of the 5-min oscillations. A two-dimensional MHD simulation that demonstrates that the photospheric 5-min oscillations can leak into the chromosphere inside small-scale vertical magnetic flux tubes was attempted by Khomenko et al. (2008). The main conclusion is that the efficiency of the energy exchange by radiation in the solar photosphere can lead to a significant reduction of the cutoff frequency and may allow for the propagation of 5-min waves vertically into the chromosphere.

One of the recent theories on the excitation mechanism for 5-min oscillations has been studied by Xiong and Deng (2009). They calculated the nonadiabatic oscillations of low- and intermediate-degree ($l = 1$–25) g4-p39-modes for the Sun. Both the thermodynamic and dynamic couplings are taken into account by using our nonlocal and time-dependent theory of convection. The results show that all the low-frequency f- and p-modes with periods $P > 5.4$ min are pulsationally unstable, while the coupling between convection and oscillations is neglected. However, when the convection coupling is taken into account, all the g- and low-frequency f- and p-modes with periods longer than ≈ 16 min (except the low-degree $p1$- modes) and the high-frequency p-modes with periods shorter than ≈ 3 min become stable, and the intermediate-frequency p-modes with period from ≈ 3 to ≈ 16 min are pulsationally unstable. The pulsation amplitude growth rates depend only on the frequency and almost do not depend on l. They achieve the maximum at $\approx 3,700 \mu$Hz (or $P \approx 270$ s). The coupling between convection and oscillations plays a key role for the stabilization of low-frequency f- and p-modes and excitation of intermediate-frequency p-modes. They proposed that the solar 5-min oscillations are not caused by any single excitation mechanism, but they are resulted from the combined effect of regular coupling between convection and oscillations and turbulent stochastic excitation. For low- and intermediate-frequency p-modes, the coupling between convection and oscillations dominates, while for high-frequency modes, stochastic excitation dominates.

Suffice it to say that the detection of 5-min oscillations in the Sun has opened up a new branch of solar physics, namely, helioseismology. The basic equations, assumptions, and other aspects of these global modes will be dealt with separately in the next chapter (Chap. 8).

7.2 Oscillations in Sunspots

Observations of oscillations in sunspots began in the late 1960s with the discovery of umbral flashes in the Ca, II, H, and K lines by Beckers and Tallant (1969). In the following years, the running penumbral waves in H_{alpha}, and the 3-min oscillations in the photosphere and chromosphere were reported by Beckers and Schwarz (1972); Bhatnagar and Tanaka (1972); Giovanelli (1972), while the important 5-min oscillations in the umbral photosphere were reported by Bhatnagar et al. (1972).

One can classify oscillations in sunspot umbras as two distinct types: 3-min and 5-min oscillations. Of these, the 3-min oscillations manifest both as velocity oscillations and as umbral flashes, being resonant modes of the sunspot, while the 5-min oscillations represent the passive response of the sunspot to forcing by the 5-min p-mode oscillations in the surrounding convection zone.

The 3-min oscillations, as mentioned above, are generally considered to be a resonant mode of the sunspot itself. Here again, two types of resonant modes are possible that have a periodicity of 3 min. They are classified as the photospheric resonance and the chromospheric resonance. The resonance at the photosphere consists of magnetoatmospheric waves that are trapped in the photosphere and subphotosphere (Uchida and Sakurai 1975). These wave modes are excited by overstable convection in the umbral subphotosphere. Another alternative is that the photospheric resonance is excited by the high-frequency components of the 5-min p-mode oscillations in the surrounding convection zone. The chromospheric resonance as compared to the photospheric resonance consists of slow magnetoacoustic waves that are nearly trapped in the chromosphere, essentially between the temperature minimum and the transition region. The excitation mechanism for these modes is typically below the photospheric resonance or by acoustic waves from the convection zone. A schematic sketch of the possible modes of oscillations in a sunspot, described by a flux tube, is presented in Fig. 7.1.

We give a brief description below of the theoretical arguments for the existence of the two types of resonances in a sunspot. Consider a compressible, inviscid, perfectly conducting gas, permeated by a uniform magnetic field $\mathbf{B}$ in the vertical direction (z-direction), which is in hydrodynamic equilibrium with uniform gravity g (in the downward z-direction). The undisturbed pressure, density, and temperature are assumed to be a function of the vertical coordinate only. The linearized magnetoatmospheric waves equations for small adiabatic perturbations can be reduced to the following pair of equations for the z-dependent amplitudes $u(z)$ and $w(z)$ of the horizontal and vertical velocities, respectively (Thomas 1983):

$$\left[v_A^2 \left(\frac{d^2}{dz^2} - k^2 \right) - c_S^2 k^2 + \omega^2 \right] u + ik \left[c_S^2 \frac{d}{dz} - g \right] w = 0 \qquad (7.1)$$

$$ik \left[c_S^2 \frac{d}{dz} - (\gamma - 1)g \right] u + \left[c_S^2 \frac{d^2}{dz^2} - \gamma g \frac{d}{dz} + \omega^2 \right] w = 0. \qquad (7.2)$$

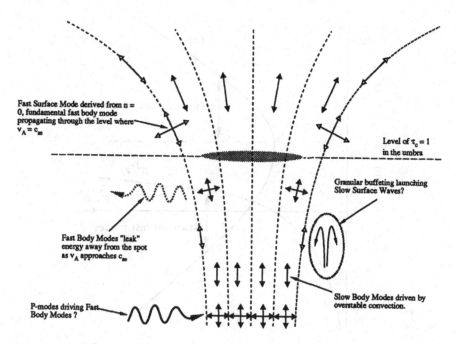

Fig. 7.1 A sketch of the various possible waves in a sunspot on the basis of a simple flux tube; from Roberts (1990)

Here, c_S is the adiabatic sound speed and v_A is the Alfvén speed, both of which vary with height z, and γ is the ratio of specific heats. Depending on the geometry being used, for example, in the case of the Cartesian coordinate system, the velocity components are written as

$$u(x,z,t) = u(z)\exp[i(kx - \omega t)], \quad w(x,z,t) = w(z)\exp[i(kx - \omega t)]. \tag{7.3}$$

For the cylindrical geometry, the velocity components are

$$u_r(r,z,t) = ku(z)J_1(kr)\exp(-i\omega t) \quad u_z(x,z,t) = -ikw(z)J_0(kr)\exp(-i\omega t), \tag{7.4}$$

where J_0 and J_1 are the Bessel functions and k is the radial wavenumber.

Once the undisturbed temperature $T(z)$ as a function of height z is specified, the undisturbed density $\rho(z)$ and pressure $p(z)$ are determined from the equation of state and the hydrostatic equations. The sound speed c_S and the Alfvén speed are respectively given by $c_S^2 = \gamma RT(z)$ and $v_A^2 = B^2/\mu\rho(z)$. The behavior of the solutions for the velocity components may be determined completely for a specified horizontal wavenumber k and frequency ω with a specified variation of the sound speed and the Alfvén speed. A schematic diagram of the variation of the sound speed with the Alfvén speed with height in a sunspot umbra which identifies the regions of trapping of the photospheric and chromospheric resonances is presented in Fig. 7.2.

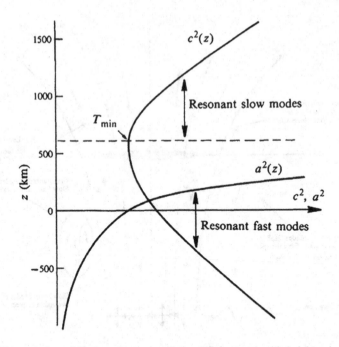

Fig. 7.2 The variation of the sound speed and the Alfvén speed with height z in a sunspot umbra. The approximate ranges of height in which the photospheric fast-mode resonance and chromospheric slow-mode resonance occur is also shown. Note the change in the notation for the sound and Alfvén speed in the figure; from Thomas (1985)

The photospheric resonance is caused by trapping of fast Alfvén speed with height up into the photosphere and upward reflection due to the increasing sound speed down into the convection zone. The compressive and magnetic forces play an important role in this case, while the buoyancy force plays a negligible role. The chromospheric resonance involves slow modes that are essentially pure acoustic waves with motions only along the vertical field lines. The resonance in this case is caused by trapping of these waves by downward reflection at the chromosphere–corona transition region due to the rapid increase in the sound speed and upward reflection at the temperature minimum, which may tend to increase the acoustic cutoff frequency with a decreasing temperature toward the temperature minimum. In most of the theoretical calculations on umbral oscillations, a purely vertical magnetic field is assumed. However, some studies on the effect of spreading magnetic field lines on the three-layer model show that the spreading magnetic field allows a greater flux of small-scale Alfvén waves up into the chromosphere (Cally 1983).

Five-minute oscillations in the sunspots are distinctly different from the 3-min oscillations in the umbra and the running penumbral waves. It was suggested by Thomas (1981) that these 5-min oscillations are the response of the sunspot to forcing by the 5-min p-mode oscillations in the surrounding atmosphere. A simple model to study the interaction of p-modes with a sunspot magnetic flux tube has been attempted by Abdelatif (1985), who found that the transmission of wave energy into the sunspot is a function of the horizontal wavenumber as compared to the tube diameter. He also found that the horizontal wavelength of the wave changed upon transmission into the magnetic flux tube. This will shift the power along the k-axis (in the $\omega - k$-plane) in the diagnostic diagram of oscillations inside the sunspot compared with that of the surroundings.

One of the important aspects of sunspots is the famous Evershed effect, which has to do with the radial motions in the sunspots. The flow occurs along arched, elevated flow channels. Recent results from the Hinode support such a scenario. Ichimoto et al. (2007) have shown that the Evershed flows in the outer penumbra have the flow velocity vector and the magnetic field well aligned, with an angle of $30°$ to the solar surface. The arched nature of the flow channels and the supersonic, field-aligned downflows in the outer penumbra are neatly reproduced in the siphon flow model by Montesinos and Thomas (1997). An excellent review on the theoretical models of sunspot structure and dynamics that deals with umbral magnetoconvection, the inner and outer penumbra, the formation and maintenance of the penumbra, magnetic field configuration in penumbra, and numerical simulations of a sunspot is available in Thomas (2010).

An excellent review on the observational aspects of oscillations in the sunspots is found in Bogdan and Judge (2006). Interested readers are advised to read the well-documented book on sunspots and starspots by Thomas and Weiss (2008).

7.3 Chromospheric Oscillations

Oscillations in the solar chromosphere were detected shortly after the observation of oscillations at the level of the photosphere (now identified as p-modes) in the 1960s. However, the understanding of chromospheric oscillations is not as complete as for photospheric ones, due in part to the inhomogeneity of the chromosphere (arising from the influence of magnetic field), which makes the observation and theoretical description more complicated. As noted by Harvey et al. (1993), very different results have been obtained, and even today, despite the wealth of data provided by several instruments, notably on board the Solar and Heliospheric Observatory (SOHO) spacecraft, differing results appear (in particular about the presence or absence of oscillations at a given height in the atmosphere) and the theoretical understanding of these oscillations remains incomplete.

At the chromospheric level, 3-min oscillations were first detected (see, e.g., Jensen and Orrall 1963), but periodicity at both 3 and 5 min was observed later on: Five-minute oscillations were detected in the chromospheric network, while 3-min oscillations were associated with internetwork regions (Dame et al. 1984). Among the remaining open questions are their very nature [acoustic, magnetoacoustic, or purely magnetic (Alfvén) waves], their attenuation with increasing height in the solar atmosphere, and their propagation through the atmosphere or their standing nature. Among the many works concerning these questions, we note Flec and Deubner (1989), who described the chromospheric oscillations as progressive waves below a magnetic height of approximately 1,000 km and standing waves above; Bocchialini and Baudin (1995), who observed propagation of waves based on wavelet analysis; Carlsson et al. (1997), who observed correlation between two lines formed at different heights and interpreted this correlation as a signature of upward propagation; Goutterbroze et al. (1999), who put a lower limit (corresponding to 50,000 K) on the existence of intensity oscillations versus height by using different lines formed at increasing temperature; Wikstol et al. (2000), who observed oscillations in the chromosphere–corona transition region (in C II and O VI lines) associated with upward, and also possibly downward, propagation (a result also obtained by McAteer et al. (2003) for waves in bright points); and, finally, the work based on a wide observational set by Judge et al. (2001), who described the chromospheric oscillations as propagating waves as a response of the chromosphere to the forcing of photospheric p-modes.

Several attempts have been made to describe these oscillations theoretically and numerically. For example, with their simulations, Carlsson and Stein (1997) have successfully described the oscillatory behavior of intensity variations in the so-called K grains. These simulations have then been widely used and compared successfully to observations. Later, Rosenthal et al. (2002) included MHD effects in the Carlsson and Stein (1997) approach; in their 2D simulation of waves in a magnetized atmosphere, they found that in regions where the field is significantly inclined to the vertical, the effect is to reflect all, or almost all, of the wave energy back downward, the altitude of the reflection being highly variable. In contrast, in the regions where the field is close to vertical, the waves that present the characteristics of a pure acoustic oscillation continue to propagate upward, channeled along the magnetic field lines. McIntosh et al. (2001) mentioned, on the basis of a comparison between different data sets, that mode mixing could occur in the chromosphere, which could explain different results from different data sets.

In principle, one can expect a multitude of waves and wave modes to be excited in the solar convection zone. In most of the convection zone, the excited waves will be predominantly acoustic in nature. When acoustic waves reach the height where the Alfvén speed is comparable to the sound speed, they undergo mode conversion, refraction, and reflection. In an inhomogeneous, dynamic chromosphere, this region of mode conversion will be very irregular and will change in time. We thus expect complex patterns of wave interactions that are highly variable in time and space. Simplified treatments imposing symmetries may lead to erroneous conclusions. However, this complex nature of the interactions and also numerical modeling of MHD waves necessitates the study of simplified geometries to examine the validity of analytic results and to build up the interpretation framework for more realistic magnetic topologies.

Rosenthal et al. (2002) made two-dimensional simulations of the propagation of waves through a number of simple field geometries in order to obtain a better insight into the effect of differing field structures on the wave speeds, amplitudes, polarizations, direction of propagation, etc. In particular, they studied oscillations in the chromospheric network and internetwork. They found that acoustic, fast-mode waves in the photosphere become mostly transverse, magnetic fast-mode waves when crossing a magnetic canopy where the field is significantly inclined to the vertical. Refraction by the rapidly increasing phase speed of the fast modes results in a total internal reflection of the waves.

Bogdan et al. (2003) reported on similar simulations in a field geometry similar to a sunspot. Four cases were studied: excitation by either a radial or a transverse sinusoidal perturbation and two magnetic field strengths, either an umbra at the bottom boundary with a field strength of 2,750 G or a weak-field case with a field strength four times smaller. In the strong-field case, the plasma β is below unity at the location of the piston and the upward-propagating waves do not cross a magnetic canopy. Because the field is not exactly vertical at the location of the piston, both longitudinal and transverse waves are excited. The longitudinal waves propagate as slow-mode, predominantly acoustic, waves along the magnetic field. The transverse waves propagate as fast-mode, predominantly magnetic, waves. These waves are not confined by the magnetic field, and they are refracted toward regions of lower Alfvén speed. They are thus turned around and they impinge on the magnetic canopy in the penumbral region. Where the wave vector forms a small angle to the field lines, the waves convert to slow waves in the lower region; where the attack angle is large, there is no mode conversion and the waves continue across the canopy as fast waves. The simulations show that wave mixing and interference are important aspects of oscillatory phenomena also in sunspots.

The solar chromosphere, which one treats as the region above the photospheric surface, is defined as the layer with continuum optical depth unity, with an extension of about 2 mm. Above the temperature minimum and up to the middle chromosphere (close to a height of about 1.5 mm), the atmosphere can be effectively regarded as nearly isothermal. In the higher layers, the observed properties of the chromosphere are strongly influenced by magnetic fields. A convenient way to parameterize the field strength is in terms of β. The $\beta = 1$ (which is not uniform with height) provides a natural separation of the atmosphere into magnetic and nonmagnetic (or weakly magnetized) regions. In the lower chromosphere and below, the magnetic field occurs structured in the form of magnetic flux tubes, which occur at the cell boundaries and constitute the well-known magnetic network. These tubes are mainly vertical and in pressure equilibrium with the outside medium. They expand upward to conserve magnetic flux. From a low filling factor ($<1\%$) in the photosphere, the tubes spread to 15% in the layers of formation of the emission features in the H and K lines of ionized calcium (at a height of 1 mm) and to 100% in the so-called magnetic canopy. The remaining quiet Sun outside the network is called the internetwork, sometimes also referred to as the cell interior.

Figure 7.3 schematically shows the structure of the chromosphere, in particular highlighting the effects of the magnetic field. The regions between the super granules are the sites of the strong network magnetic fields and also where mottles/spicules occur. The network patches are connected by essentially horizontal field lines. The internetwork region is weakly magnetized below the canopy.

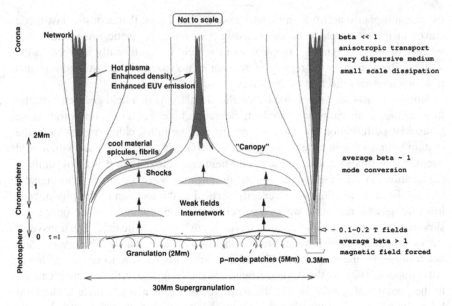

Fig. 7.3 Structure of the chromosphere, in particular highlighting the effects of the magnetic field. The regions between the super granules are the sites of the strong network magnetic fields. The network patches are connected by essentially horizontal field lines, the internetwork region is weakly magnetized below the canopy, whereas above it the chromosphere is magnetized; from Judge (2006)

Quantitative studies of wave propagation in magnetically structured and gravitationally stratified atmospheres help to identify various physical mechanisms that contribute to the dynamics of the magnetic network on the Sun, and to develop diagnostic tools for the helioseismic exploration of such atmospheres. Magnetic fields play an important role in wave generation and propagation. A simple model of the network element consisting of individual flux tubes for dynamical simulations is presented in Fig. 7.4.

Vigeesh et al. (2009) carried out a number of numerical simulations of wave propagation in a two-dimensional gravitationally stratified atmosphere consisting of individual magnetic flux concentrations representative of solar magnetic network elements of different field strengths. While the magnetic field in the internetwork regions of the quiet Sun is mainly shaped by the convective-granular flow with a predominance of horizontal fields and a rare occurrence of flux concentrations surpassing 1 kg, the magnetic network shows plenty of flux concentrations at or surpassing this limit, with a typical horizontal size scale in the low photosphere of 100 km. These magnetic elements appear as bright points in G-band images near the disk center and can be modeled well as magnetic flux tubes and flux sheets. The results of the simulations for the temperature and velocity are shown in Figs. 7.5 and 7.6, respectively. The results show that for transverse, impulsive excitation, flux tubes/ sheets with strong fields are more efficient in providing acoustic flux to the

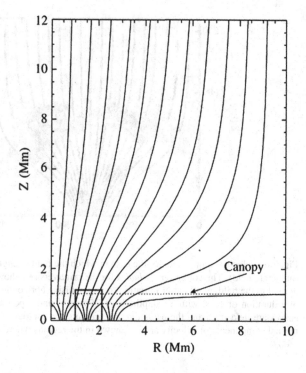

Fig. 7.4 Model of a network element consisting of individual flux tubes separated at the photospheric surface by a distance of 1,000 km that merge at a height of about 600 km. The *box* corresponds to the domain taken for dynamical simulations; from Hasan et al. (2005)

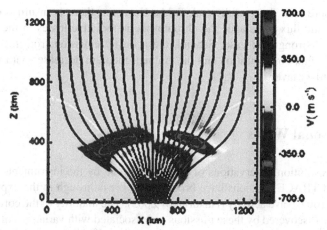

Fig. 7.5 Temperature perturbations for the case in which the field strength at the axis at $z = 0$ is 800 G (moderate field). The colors (*gray shades* for the print version) show the temperature perturbations at 40 s after initiation of an impulsive horizontal motion at the $z = 0$ boundary of a duration of 12 s, with an amplitude of 750 ms^{-1} and a period of $P = 24$ s. The *thin black curves* are field lines, and the *white curve* represents the contour of $\beta = 1$; from Vigeesh et al. (2009)

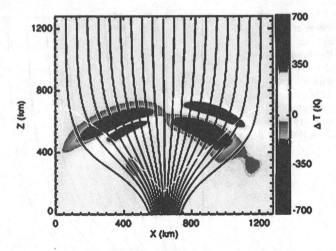

Fig. 7.6 Velocity components for the case in which the field strength at the axis at $z = 0$ is 800 G (moderate field). The colors (*gray shades* for the print version) show the velocity components V_n, normal to the field, at 40 s after initiation of an impulsive horizontal motion at the $z = 0$ boundary of a duration of 12 s, with an amplitude of $750\,\mathrm{m\,s}^{-1}$ and a period of $P = 24$ s. The *thin black curves* are field lines and the *white curve* represents the contour of $\beta = 1$. The field-aligned and normal components of velocity are not shown in the regions where $B < 50\,\mathrm{G}$; from Vigeesh et al. (2009)

chromosphere than those with weak fields. However, there is insufficient energy in the acoustic flux to balance the chromospheric radiative losses in the network, even for the strong-field case. Second, the acoustic emission from the interface between the flux concentration and the ambient medium decreases with the width of the boundary layer.

7.4   Coronal Waves

The high-resolution observations of the solar corona by the instruments on board SOHO and TRACE missions have brought us a breakthrough in the experimental study of coronal wave activity (Nakariakov 2003). The majority of the coronal wave phenomena discovered by these missions are associated with variations of the EUV emissions produced by coronal plasmas. The characteristic speeds of the variations are found to be about several minutes. Observationally determined properties of various coronal waves and oscillations are quite different, allowing us to distinguish between several kinds of phenomena. In particular, kink and longitudinal modes are confidently interpreted in the data. The modes of other types, in particular sausage and torsional, predicted by theory, have not been identified in EUV data. However, the sausage modes are believed to be routinely observed in the radio band.

Unresolved torsional waves can be responsible for the modulation of nonthermal broadening of coronal emission lines. The observational discovery of coronal waves reinforced wave-based theories of coronal heating and led to the creation of a new branch of coronal physics: MHD coronal seismology, first proposed as a theoretical possibility by Uchida (1970) and Roberts (1984). Measuring the properties of MHD waves and oscillations (periods, wavelengths, amplitudes, temporal and spatial signatures, characteristic scenarios of the wave evolution), combined with theoretical modeling of the wave phenomena (dispersion relations, evolutionary equations, etc.), we can determine values of the mean parameters of the corona, such as the magnetic field strength and transport coefficients. The first practical implementation of this method was made by Nakariakov (1999, 2001).

Theoretical aspects of MHD waves in the solar coronal plasma have been investigated for decades. There have also been several reports on the existence of MHD phenomena in the solar corona (Rosenberg 1970; Aschwanden 1987; Svestka et al. 1982; Koutchmy et al. 1983; McKenzie et al. 1987). With the spatial detection of oscillations by the Transition Region and Coronal Explorer (TRACE) spacecraft, theories on MHD waves have taken on a new vigor. It has now been proven possible to systematically study coronal loop oscillations and their decay (Schrijver et al. 2002; Aschwanden et al. 2002). Oscillations in hot loops have also been very recently detected by the Solar Ultraviolet Measurements of Emitted Radiation (SUMER) spectrometer on the Solar and Heliospheric Observatory (SOHO) (Kliem et al. 2002; Wang et al. 2002). The theory of coronal loop oscillations has been reviewed (Roberts 1991; Roberts 2000; Nakariakov 2001; Goossens et al. 2002; Roberts and Nakariakov 2003). However, it is evident that the subject is developing apace, led by the recent observational discoveries that have prompted a reexamination of theoretical aspects.

Acoustic waves in coronal loops: Slow magnetosonic waves (acoustic waves) are an abundant feature of the coronal wave activity, known from observations such as SOHO/EIT, UVCS, SUMER, and TRACE. These modes are longitudinal in nature, perturbing the density of the plasma and the parallel component of the velocity. Both propagating and standing waves are observed.

Propagating longitudinal waves: With imaging telescopes, propagating longi-tudinal waves are observed in both open and closed coronal magnetic structures. The standard observational technique is the use of the stroboscopic method: The emission intensities along a chosen path, taken in different instants of time, are laid side by side to form a time-distance map. Diagonal stripes of these maps exhibit disturbances, which change their position in time and, consequently, propagate along the path. This method allows the determination of periods (or wavelengths), relative amplitudes, and projected propagation speeds. The first observational detection of longitudinal waves came from analyzing the polarized brightness (density) fluctuations. The fluctuations with periods of about 9 min were detected in coronal holes at a height of about $1.9\,R_\odot$ by Ofman et al. (1997, 1998) using the white light channel of the SOHO/UVCS. Theoretical models of the propagation of longitudinal waves in stratified coronal structures (Ofman et al. 1999, 2000;

Nakariakov et al. 2000a), describe the evolution of the wave shape and amplitude with the distance along the structure s in terms of the extended Burgers equation,

$$\frac{\partial A}{\partial s} - \alpha_1 A - \alpha_2 \frac{\partial^2 A}{\partial \xi^2} + \alpha_3 A \frac{\partial A}{\partial \xi} = 0, \tag{7.5}$$

where the coefficients α_1, α_2, and α_3 are in general functions of s and describe α_1—the effects of stratification, radiative losses, and heating, α_2—dissipation by thermal conductivity, and viscosity, and α_3—nonlinearity; and the $\xi = s - C_s t$ is the running coordinate.

Solutions of the above equation are in satisfactory agreement with the observed evolution of the wave amplitude. Also, full MHD 2D numerical modeling of these waves gives similar results. The energy carried and deposited by the observed waves is certainly insufficient for heating of the loop. However, Tskilauri and Nakariakov (2001) has shown that wide spectrum slow magnetoacoustic waves, consistent with currently available observations in the low-frequency part of the spectrum, can provide the rate of heat dissipation sufficient to heat the loop.

Kink oscillations: Movies created with the use of coronal imaging data show fast-decaying quasi-periodic displacement of loops, often responding to an energy release nearby, in a form of a flare or eruption (flare-generated oscillations). Analysis of 26 oscillating loops with lengths of 74–582 mm, observed in EUV with TRACE (Aschwanden et al. 2002), yielded periods of 2.3–10.8 min, which are different for different loops. It is known that coronal loops are anchored in the dense plasma of the photosphere, so it is reasonable to assume that any motions in the corona are effectively zero at the base of a loop. The observed properties of these oscillations can be interpreted in terms of a kink fast magnetosonic mode (Edwin and Roberts 1983). The first observation of kink oscillations was after the flare on July 14, 1998, at 12.55 UT. The oscillation was identified as a global mode, with the maximum displacement situated near the loop apex and the nodes near the foot points. Using the above theory, Nakariakov (2001) estimated the magnetic field in an oscillating loop as 13 ± 9 G. A typical sketch of the loop oscillation is presented in Fig. 7.7. The effect of uniform flows on kink oscillations was studied by Satya Narayanan et al. (2004). See also the review article by Satya Narayanan and Ramesh (2005). We briefly describe their results: Assume the plasma $\beta \ll 1$. The pressure balance condition is given by

$$p_0 + \frac{B_0^2}{2\mu} = p_e + \frac{B_e^2}{2\mu}. \tag{7.6}$$

For $\alpha = \rho_e/\rho_0$, $\varepsilon = U_0/c_{A1}$, $x = \omega/kc_{A1}$, and low-β plasma, it can be shown that

$$m_0 = k[1 - (x - \varepsilon)^2]^{1/2} = m_0^* \tag{7.7}$$

$$m_e = k[1 - \alpha x^2]^{1/2} = m_e^*. \tag{7.8}$$

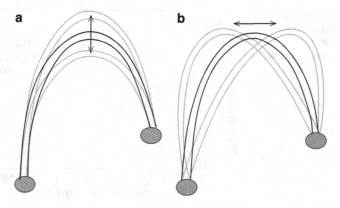

Fig. 7.7 A loop oscillation; from Nakariakov and Ofman (2001)

Fig. 7.8 The behavior of $F_m(ka)$ for different values of ka; from Satya Narayanan et al. (2004)

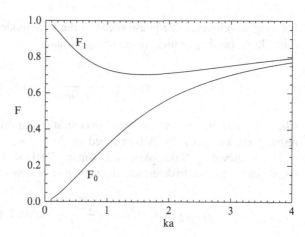

The dispersion relation for low-β plasma with flow can be written as

$$[(x-\varepsilon)^2 - 1](1 - \alpha x^2)^{1/2} + \alpha(x^2 - 1)[1 - (x-\varepsilon)^2]^{1/2}F(m_0^*, m_e^*, a) = 0 \quad (7.9)$$

$$F(m_0^*, m_e^*, a) = \frac{K_m(m_e^* a)I_m'(m_0^* a)}{K_m'(m_e^* a)I_m(m_0^* a)}. \quad (7.10)$$

The behavior of $F_m(ka)$ for different values of ka is shown in Fig. 7.8. The above relation, Eq. (7.10), is highly transcendental and will have to be solved numerically. However, for $ka \ll 1$, one can show that $F(m_0^*, m_e^*, a) \approx 1$, so that the dispersion relation would reduce to

$$[(x-\varepsilon)^2 - 1](1 - \alpha x^2)^{1/2} + \alpha(x^2 - 1)[1 - (x-\varepsilon)^2]^{1/2} = 0. \quad (7.11)$$

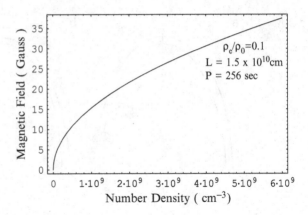

Fig. 7.9 The variation of the magnetic field for different coronal parameters; from Satya Narayanan et al. (2004)

For long wavelengths, the phase speed of the kink mode is about equal to the so-called kink speed c_K, which, in the low-β plasma, is

$$c_K \approx \left[\frac{2}{1 + n_e/n_0} \right]^{1/2} c_{A1}, \tag{7.12}$$

where n_0 and n_e are the plasma concentrations inside and outside the loop, respectively, and c_{A1} is the Alfvén speed inside the loop.

It was shown by Nakariakov and Ofman (2001) that the formula for the kink speed can be utilized to determine the magnetic field as follows:

$$B_0 = (4\pi\rho_0)^{1/2} c_{A1} = \frac{\sqrt{2}\pi^{3/2}L}{P} \sqrt{\rho_0(1 + \rho_e/\rho_0)}. \tag{7.13}$$

Figure 7.9 presents the variation of the magnetic field for different coronal parameters.

Sausage oscillations: Modulated coronal radio emissions that have periodicity in the range 0.5–5 s have been interpreted in terms of a fast magnetoacoustic mode, the sausage mode, associated with the perturbations of the loop cross section and plasma concentration by Nakariakov et al. (2003). Quasi-periodic pulsations of shorter periods (0.5–10 s) may be associated with sausage modes of higher spatial harmonics Roberts et al. (1984); Nakariakov and Ofman (2001). There have been quasi-periodic pulsations in the periods 14–17 s, which oscillate in phase at a loop apex, and its foot points, which have been observed at radio wavelengths. These modes have a maximum magnetic field perturbation at the loop apex and nodes and at the foot points. Dependence of the cutoff wavenumber of the sausage mode is depicted in Fig. 7.10.

The dispersion relation for magnetoacoustic waves in cylindrical magnetic flux tubes has many types of long-wavelength solutions in the fast-mode branch

Fig. 7.10 Dependence of the cutoff wavenumber of the sausage mode; from Nakariakov and Ofman (2001)

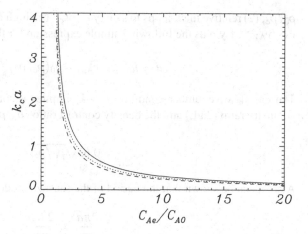

$(n = 0, 1, 2, \ldots)$, with the lowest case called the sausage $(n = 0)$ and kink mode $(n = 1)$. Kink mode solutions extend all the way to the long-wavelength limit $(ka \to 0)$, while the sausage mode has a cutoff at a phase speed:

$$v_{\text{ph}} = v_{\text{A2}}, \tag{7.14}$$

which has no solutions for wavenumbers $ka < k_c a$.

The cutoff wavenumber k_c is given by

$$k_c = \left[\frac{(c_s^2 + v_{\text{A1}}^2)(v_{\text{A2}}^2 - c_T^2)}{(v_{\text{A2}}^2 - v_{\text{A1}}^2)(v_{\text{A2}}^2 - c_s^2)} \right]^{1/2} \left(\frac{j_0}{a} \right). \tag{7.15}$$

Under coronal conditions, the sound speed $c_s \approx 150$–260 km/s and Alfvén speed is $v_A \approx$ a few hundred km/s. Therefore,

$$c_s \ll v_A. \tag{7.16}$$

Here tube speed is similar to the sound speed,

$$c_T \approx c_s \tag{7.17}$$

$$k_c \approx \left(\frac{j_0}{a} \right) \frac{1}{[(v_{\text{A2}}/v_{\text{A1}})^2 - 1]^{1/2}}. \tag{7.18}$$

For a typical density ratio in the solar corona (0.1–0.5), the cutoff wavenumber $k_c a$ falls in the range $0.8 \le k_c a \le 2.4$. Therefore, the long-wavelength sausage mode oscillation is completely suppressed for the slender loops. The occurrence of global sausage modes therefore requires special conditions: (1) a very high density contrast

ρ_0/ρ_e; (2) relative thick loops to satisfy $k > k_c$. The high density ratio $\rho_0/\rho_e \gg 1$ or $v_{A2}/v_{A1} \gg 1$ yields the following simple expression for the cutoff wavenumber k_c:

$$k_c a \approx j_0(v_{A1}/v_{A2}) = j_0(\rho_e/\rho_0)^{1/2}. \tag{7.19}$$

The cutoff wavenumber condition $k > k_c$ implies a constraint between the loop geometry ratio (2a/L) and the density contrast ratio (ρ_e/ρ_0), which turns out to be

$$\frac{L}{2a} \approx 0.65\sqrt{\rho_0/\rho_e}. \tag{7.20}$$

Also, it can be shown that the period of the sausage mode satisfies the condition

$$P_{\text{saus}} < \frac{2\pi a}{j_0 v_{A1}} \approx \frac{2.62a}{v_{A1}}. \tag{7.21}$$

Observations of radio burst emission from the "disturbed" Sun at meter wavelength seem to provide the bulk of available evidence for coronal oscillations (Aschwanden et al. 1994). One of the important reasons for this is the high temporal resolution with which data can be obtained. Also, according to Aschwanden (1987), the favorable conditions for MHD oscillations occur mainly in the upper part of the corona. Since the radio emissions observed in the meter wavelength range originate typically at heights $> 0.2\,R_\odot$ above the solar photosphere, they play a useful role in this connection. We recently analyzed the data obtained with the Gauribidanur radioheliograph (Ramesh et al. 1998) and Mauritius radio telescope (Ramesh et al. 1998) for quasi-periodic emission from the solar corona . The theory of MHD oscillations was used to determine the Alfvén speed and magnetic field. The estimated values are in the range 800–1,200 km/s and 3–30 G, respectively (Ramesh et al. 2003, 2004, 2005).

It is a common belief that microwave bursts are generated by the gyro synchrotron emission, which is very sensitive to the magnetic field in the radio source. Causes of microwave flux pulsations with periods $P \approx 1$–20 s are believed to be some kind of magnetic field fluctuations that modulate the gyro synchrotron radiation, leading to acceleration of particles. The observational proof of the existence of a global sausage mode should be based upon the determination of the oscillation period, the longitudinal and transverse size of the magnetic loop, and the spatial distribution of the oscillation amplitude along the loop. A good candidate for such a proof is a solar flare, which happened on January 12, 2000, and was observed by Nobeyama Radioheliograph, Japan, at two frequencies. Details of the observations can be found in Nakariakov et al. (2003). The time profile and the Fourier power spectra of the pulsations are presented in Fig. 7.11. It is clear from the figure that these variations are quasi-periodic in nature and may be interpreted in terms of sausage oscillations.

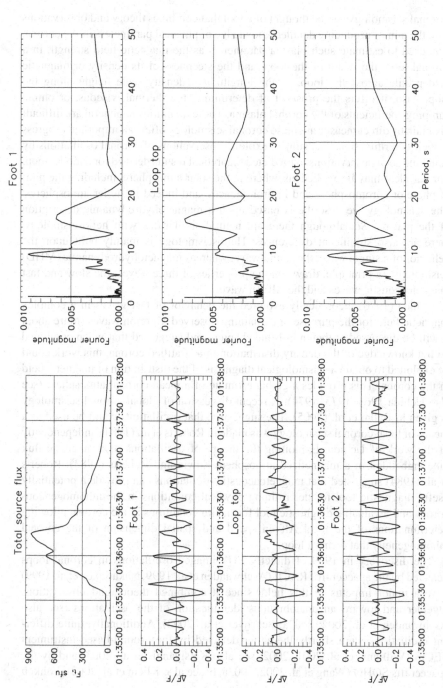

Fig. 7.11 The time profile and the Fourier power spectra of the pulsations; from Nakariakov et al. (2003)

7.5   Coronal Seismology

Coronal seismology can be thought of a tool that combines theory and observations with the aim of producing detailed knowledge of physical parameters in the corona. It is used to estimate such plasma parameters as the magnetic field strength in a coronal loop, the width of the loop, and the steepness of its density or magnetic field profile across the loop, or the longitudinal density scale height along the loop. Also, it offers the prospect of determining the thermal, viscous, or ohmic damping coefficients of the coronal plasma. These quantities in general are difficult to obtain by direct measurement, so coronal seismology offers an important progress in their determination. The subject exploits observations of coronal oscillations by matching such observations to predicted theoretical results derived for a model loop. Similar ideas may be applied elsewhere in the solar atmosphere, including the photosphere or chromosphere and in prominences, and indeed in stellar atmospheres. The seismology we describe is based upon a magnetohydrodynamic description of the plasma. So, although there are natural similarities with helioseismology, there are also significant differences: Helioseismology is mainly built upon the behavior of a single wave, the sound wave, whereas magnetohydrodynamic (MHD) seismology, in principle, draws on the properties of three waves: the slow and fast magnetoacoustic waves and the Alfvén wave.

Uchida (1970) theoretically explored the behavior of fast waves in a complex magnetic field for the purpose of explaining observed Moreton waves (more about it will be discussed at the end of this chapter) and suggested that when combined with a knowledge of the density distribution in a stratified corona, this work could be exploited to obtain a seismological diagnosis of the distribution of magnetic field in the corona. This provides a global estimate of a mean coronal atmosphere (see also Uchida 1968, 1973, 1974). A recent discussion of global coronal seismology is given in Ballai et al. (2005). The suggestion that oscillations could be used as a means of local coronal seismology was made in Roberts et al. (1984), independently of the work of Uchida. Exploiting the theory of oscillations of a magnetic flux tube embedded in a magnetized atmosphere Edwin and Roberts (1983); Roberts et al. (1984) suggested that magnetoacoustic oscillations can provide a potentially useful diagnostic tool for determining physical conditions in the inhomogeneous corona and a valuable diagnostic tool for in situ conditions in the corona, allowing determinations of the local Alfvén speed and spatial dimension of the coronal inhomogeneity that forms a loop.

The discovery in 1999 of directly EUV-imaged oscillations in coronal loops detected by the spacecraft TRACE (Aschwanden et al. 1999; Nakariakov et al. 1999) gave a major impetus to this field, since it combined theory and observations together and toward an unambiguous identification of the oscillations (see also Aschwanden et al. 2002; 2002a; Schrijver et al. 2002). Additionally, quite different oscillations were shortly thereafter detected by the ground-based instrument SECIS (Williams et al. 2001, 2002; see also Katsiyannis et al. 2003) and by the spacecrafts SOHO (Wang et al. 2002, 2003a, b; see also Kliem et al. 2002), Yohkoh

(Mariska 2005, 2006) and RHESSI (Foullon et al. 2005). Taken together with the long history of spatially unresolved radio observations of oscillatory phenomena in the corona (see Aschwanden 1987), and the detection of oscillations in coronal polar plumes (Ofman et al. 1997; De Forest and Gurman 1998) and also in prominences (Oliver and Ballester 2002; Diaz and Ballester 2003), we see the development of coronal seismology as a rich subject.

It is interesting to note that applications of coronal seismology are spreading; by coronal seismology, we deal with the technique of combining observations with magnetohydrodynamic wave theory, which of course can also be applied for other astrophysical objects outside the corona (with perhaps only minor modifications). In this regard, it is interesting to note the new applications to spicules (Kukhianidze et al. 2006; Singh and Dwivedi 2007; Zaqarishvili et al. 2007) and indeed to loops in stellar coronae (Mitra-Kraev et al.2005; Mathiridakis et al. 2006).

Following the direct imaging of oscillating loops by TRACE, the theoretical framework introduced through the dispersion diagram of Edwin and Roberts (1983) was used as a means of interpreting the observations (Roberts et al. 1984); the topic was reviewed in Roberts (2000), where, in addition, various issues then judged problematic were outlined. Here we take stock of the considerable progress made in the field since the observational discoveries of 1999. Reviews of theoretical aspects are also given in Roberts (2002, 2004); Roberts and Nakariakov (2003), Goossens et al. (2005), and Goossens et al. (2006). Observations have been reviewed in Wang (2004, 2006) and Banerjee et al. (2007). Broad overviews are given in Aschwanden (2004); Nakariakov and Verwichte (2005); De Moortel (2005) and Nakariakov (2007).

We mentioned in the previous section on coronal waves that the kink oscillations may be used as a diagnostic for determining the magnetic field of the solar corona. The magnetic field of the solar corona has been determined using the metric radio observations of a transient, quasi-periodic, type IV burst emission from the solar corona following the hard X-ray/Hα flare of November 23, 2000. The radio event lasted for about 121 s, and the measured mean period was 14.7 ± 2.5 s. The source region of the observed radio emission was found to be located at a height of 0.18 ± 0.03 R$_\odot$ above the solar photosphere. The Alfvén speed (v_A) at that location was estimated to be $1,185 \pm 181$ km s^{-1}. Combined with the plasma density corresponding to the observing frequency (109 MHz), this gives a magnetic field (B) of 7.2 ± 1.1 G for the above region. The speed of the disturbance propagating through the solar atmosphere in the aftermath of the flare was estimated to be ≤ 755 km s^{-1} Ramesh et al. (2004).

The radio data were obtained with the Gauribidanur radioheliograph (GRH) operating near Bangalore, India (see Ramesh et al. 1998 for details on the instrument), on November 23, 2000. The observing frequency was 109 MHz. The time resolution and bandwidth used were 128 ms and 1 MHz, respectively. Figure 7.12 shows the temporal evolution of the whole Sun flux observed with the GRH during the interval 05:33–05:43 UT on that day. One can clearly notice two phases of strong emission above the background in Fig. 7.12: (1) a couple of bursts during the period 05:35:52–05:36:30 UT, and (2) quasi-periodic emission in the interval

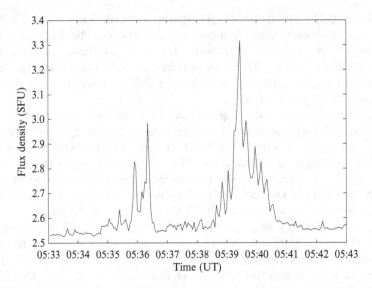

Fig. 7.12 Time profile of the radio emission from the solar corona observed with the GRH at 109 MHz on November 23, 2000, during the time interval 05:33–05:43 UT. The transient emission in the interval 05:35:52–05:36:30 UT is a type III radio burst. The quasi-periodic emission during the period 05:38:37–05:40:37 UT is a stationary type IV burst; from Ramesh et al. (2004)

05:38:37–05:40:37 UT. The estimated period of the latter from the time interval between the adjacent peaks and the average value is $14.7 \pm 2.5\,$s. The observed peak flux density was ≈ 3.3 SFU (SFU = solar flux unit = $10^{-22}\,\mathrm{W\,m^{-2}\,Hz^{-1}}$). The above radio events were closely associated with a 1F class Hα flare (05:36–06:15 UT with peak at 05:48 UT) from AR 9238 located at S26W40, and a GOES C5.4 class X-ray flare (05:34–06:17 UT with peak at 05:47 UT, Solar Geophysical Data, May 2001). Figure 7.13 shows an EIT $195\,A°$ running difference image (Bruckner et al. 1995) obtained on November 23, 2000, at 05:47 UT, by subtracting the previous image (05:35 UT).

The half-power contours of the radioheliogram obtained with the GRH during various stages of the quasi-periodic emission described above are shown in Fig. 7.14. One can notice that their centroids are distributed in the close vicinity of the flare site. A comparison with the GRH observations indicates that the first transient event around 05:36 UT in the latter (Fig. 7.12) corresponds to the above burst. The diffuse emission toward the end of the type III event and the quasi-periodic emission in the second phase of the GRH observations (Fig. 7.12) correspond to it.

The metric type IV radio burst emission from the solar corona sometimes shows periodic or quasi-periodic fluctuations superimposed on a background continuum (see Roberts et al. 1984 and the references therein). The periodicity is considered to be because of a modulation of the radio emissivity of the electrons trapped inside the associated coronal loop structure by sausage oscillations (symmetric oscillations of

Fig. 7.13 SOHO-EIT 195 $A°$ running difference image obtained on November 23, 2000, at 05:47 UT, by subtracting the previous (05:35 UT) image. Solar north is straight up and east is to the left. The *dark open circle* indicates the solar limb. The bright emission in the southwest quadrant is the flaring region; from Ramesh et al. (2004)

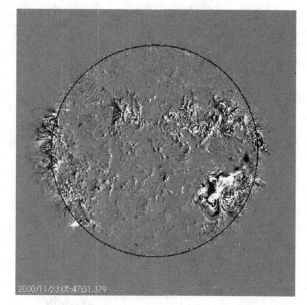

a coronal loop with its central axis remaining undisturbed) of the latter, which in turn is due to a disturbance propagating through the solar atmosphere in the aftermath of a flare (Roberts et al. 1983). A periodic injection of fast electron beams into the coronal loop as the cause for the oscillations can be ruled out in the present case since the dynamic spectrum of the type IV burst emission in Fig. 7.14 doesn't show any fine structure, as is expected (Aurass et al. 2003; Zlotnik et al. 2003a,b). Note that when oscillations develop in a coronal loop, the magnitude of the magnetic field strength and the mirror ratio in the trap are modulated. This in turn changes both the energy spectrum and the number of trapped particles. Therefore, for any generation mechanism, the radio emission flux density will be modulated with the period (p) of the oscillations. The latter has been calculated to be (see Aschwanden et al. 1999) and the references therein)

$$p = 2 \times 10^{-11} \frac{a \sqrt{N_e}}{B} \quad \text{s}, \tag{7.22}$$

where a (cm) is the radius of the coronal loop, B (G) is the associated coronal magnetic field, and N_e (cm^{-3}) is the plasma density corresponding to the observing frequency. The present observations were carried out at a frequency of 109 MHz, and therefore $N_e = 1.47 \times 10^8$ cm^{-3}. Ofman and Aschwanden (2002) have reported the parameters of coronal loops in extreme ultraviolet using the data obtained with the TRACE instrument (Handy et al. 1998). According to them, the average width of the oscillating loops is 8.7 $\pm$ 2.8 Mm. Substituting for the various values in Eq. (7.22), we get $B = 7.2 \pm 1.1$ G. Having known B and N_e, one can now

GAURIBIDANUR RADIO HELIOGRAM - 109 MHz

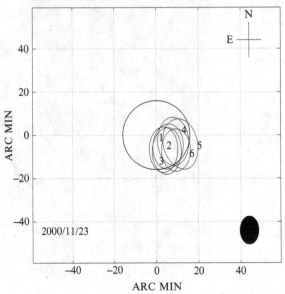

Fig. 7.14 Radioheliogram obtained with the GRH at 05:38:47 (1), 05:39:07 (2), 05:39:27 (3), 05:39:47 (4), 05:40:07 (5), and 05:40:27 UT (6), on November 23, 2000. Note that only the half-power contours are shown here for a better visualization of the event and to establish the spatial correspondence with the associated activity at other wavelengths. A comparison with Fig. 7.13 indicates that the radio sources are located in close vicinity of the flare site. The estimated peak radio brightness temperature is $\approx 1.57 \times 10^7$ K and corresponds to the image obtained at 05:39:27 UT. The *open circle* at the center is the solar limb. The size of the GRH beam at 109 MHz is indicated at the lower right corner; from Ramesh et al. (2004)

calculate the Alfvén speed (v_A) at the location from where the observed radio emission originated. From the definition of Alfvén velocity, we have

$$v_A = \frac{B}{\sqrt{4\pi M N_e}} \quad \text{cm s}^{-1}, \tag{7.23}$$

where $M = 2 \times 10^{-24}$ g is the mass ascribed to each electron in the coronal plasma (includes 10% He). The different values in Eq. (7.23) yield $v_A = 1185 \pm 181\,\text{km}\,\text{s}^{-1}$. Assuming that the disturbance responsible for the triggering of the oscillations was generated at the same time as the onset of the hard X-ray flare (i.e., at 05:35:51 UT), the altitude of the plasma level from where the observed 109-MHz radio emission originated was calculated [using the relationship between the radius of the coronal loop, duration of the quasi-periodic radio emission, and the delay in the onset of the latter from that of the hard X-ray flare (see Fig. 8 of Roberts et al.1984)], and the value is $0.18 \pm 0.03\,R_\odot$.

It is interesting to note that though the quasi-periodic type IV emission under study was preceded by a group of type III bursts, the latter did not exhibit a similar phenomenon. The quasi-periodicity in some of the type III radio burst groups (Aschwanden et al. 1994) is due to a modulation of the acceleration of the electrons released during the associated flare by a disturbance propagating through the solar atmosphere in the aftermath of the latter (see Ramesh et al. 2003, 2004 and the references therein). An absence of the above phenomenon in the present case suggests that the disturbance might have been weak to modulate the electron acceleration process. The low value of the flux density of the observed type III burst (Fig. 1; see Suzuki and Dulk (1985) for characteristics of the metric type III bursts) and the strength of the associated X-ray and Hα flare (the latter in particular) also indicate the same. The absence of type II radio bursts, which are due to plasma emission from the electrons accelerated by a magnetohydrodynamic (MHD) shock in the solar corona, strengthens the above view. Note that for the development of the latter, the speed of the disturbance must be greater than the characteristic Alfvén speed in the medium. Since metric type II radio bursts are usually observed in between the type III and IV emissions (Wild et al. 1963), the disturbance must have traveled a distance of $\leq$ 0.18 $R_\odot$ (the altitude of the source region of the quasi-periodic type IV burst in the present case) in the time interval between the onset of the hard X-ray flare and the quasi-periodic emission ($\approx$166 s). This implies that its speed must have definitely been $\leq$ 755 km s^{-1}. This is well below the Alfvén speed inside the coronal loop, derived earlier. Since global sausage-type oscillations are expected only in dense loops (Aschwanden et al. 2004), the density of the medium external to the coronal loop must be smaller than that inside. In the low-β corona, the thermal pressure is much smaller than the magnetic pressure. So one can assume almost identical magnetic field strengths both inside and outside the loop. These imply that the Alfvén speed outside the loop must be comparatively greater (refer to Eq. (7.23)). But the estimated speed of the flare- generated disturbance in the present case is sub-Alfvénic even inside the loop. Hence, the absence of a type II burst. According to Smerd (1964), though a type II burst requires fewer electrons compared to a type IV burst, they should be of higher energy. Ramesh et al. (2004) showed quantitatively that both the metric type II and quasi-periodic type III radio burst emissions are driven by the same disturbance generated in the aftermath of a flare. This particular last observational evidence, together with the above arguments, clearly explains the reason for the absence of quasi-periodicity in the type III radio burst group in the present case.

Gauribidanur observed circularly polarized emission from the solar corona at 77 MHz during the periods 11–18 August 2006, 23–29 August 2006, and 16–22 May 2007 in the minimum phase between sunspot cycles 23 and 24. The observations were carried out with the east-west one-dimensional radio polarimeter at the Gauribidanur observatory located about 100 km north of Bangalore. Two-dimensional imaging observations at 77 MHz during the same period with the radioheliograph at the same observatory revealed that the emission region co-rotated with the Sun during the aforementioned three periods. Their rotation rates, close to the central meridian on the Sun, are $4'.6$, $5'.2$, and $4'.9 \pm 0'.5$ per day, respectively.

We derived the radial distance of the region from the above observed rotation rates, and the corresponding values are $\approx 1.24 \pm 0.03\, R_\odot$ (11–18 August 2006), $\approx 1.40 \pm 0.03\, R_\odot$ (23–29 August 2006), $\approx 1.32 \pm 0.03\, R_\odot$ (16–22 May 2007). The estimated lower limits for the magnetic field strength at the above radial distances and periods are $\approx 1.1, 0.6$, and $0.9\, \mathrm{G}$, respectively (Ramesh et al. 2011).

7.6   Coronal Heating Due to Waves

Heating of the solar atmosphere, in particular the corona, is one of the outstanding problems that have intrigued scientists for several decades. Several theories and mechanisms have been proposed to explain the phenomenon of million-degree-hot corona. However, to this day, it is hard to say that the phenomenon is completely understood. Let's briefly go through some of the mechanisms to have an understanding of the complex nature of this heating. The term "heating mechanism" comprises three physical aspects: the generation of a carrier of mechanical energy, the transport of mechanical energy into the chromosphere and corona, and the dissipation of the energy in these layers. Table 7.1 (Erdelyi 2008) shows the various proposed energy carriers, which can be classified into two main categories: hydrodynamic and magnetic mechanisms. The magnetic mechanisms can be subdivided further into wave- or AC-mechanisms and current sheet- or DC-mechanisms (Ulmschneider 2003).

In order to explain the solar (and stellar) atmospheric heating, mechanisms and models have to provide tools that result in a steady supply of energy that is not necessarily steady. Random energy releases that produce a statistically averaged steady state are allowed to balance the atmospheric (chromospheric and coronal) energy losses, and these models have become more viable (MendozaBriceno et al.

Table 7.1 Various possible heating mechanisms

Energy carrier	Dissipation mechanism
Hydrodynamic heating mechanisms	
Acoustic waves ($P < P_{\text{acoustic-cutoff}}$)	Shock dissipation
Pulsation waves ($P > P_{\text{acoustic-cutoff}}$)	Shock dissipation
Magnetic heating mechanisms	
Wave mechanisms (AC) alternating current	
Slow waves	Shock damping, resonant absorption
Longitudinal MHD tube waves	
Fast MHD waves	Landau damping
Alfvén waves (transverse, torsional)	Mode coupling, resistive heating
	Phase mixing, compressional viscous
	Heating, turbulent heating, Landau damping
	Resonant absorption
Direct current (DC) mechanisms	
Current sheets	Reconnection

Fig. 7.15 Acoustic waves and heating by shocks; from Ulmschneider (2003)

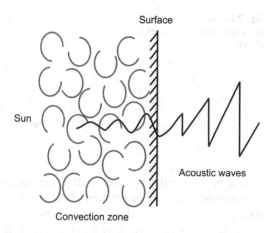

2002, 2003, 2005, 2006). Testing a specific heating mechanism observationally may be rather difficult because several mechanisms may operate at the same time. Ultimate magnetic dissipation occurs on very small spatial scales, sometimes of the order of a few hundred meters, that even with current high-spatial-resolution satellite techniques may not (and will not for a while!) be resolved. A distinguished signature of a specific heating mechanism could be obliterated during the thermalization of the input energy. However, one should instead predict the macroscopic consequences of a specific favored heating mechanism (Cargil 1993) and confirm these signatures by observations (Aschwanden 2003).

One of the predictions could be the generated flows (Ballai et al. 1998) or specific spectral line profiles or line broadenings (Erdelyi et al.1998; Taroyan and Erdelyi 2008). Without contradicting observations, it is usually not very hard to come up with a theory that generates and drives an energy carrier. The most obvious candidate is the magneto-convection right underneath the surface of the Sun.

The heating mechanisms in the solar atmosphere can be classified based on whether they involve magnetism or not. For magnetic-free regions (e.g., in the chromosphere of the quiet Sun), one can suggest a heating mechanism that yields within the framework of hydrodynamics. Such heating theories can be classified as hydrodynamic heating. Examples of hydrodynamic heating are, among others, for instance, acoustic waves and pulsations. A sketch of the heating due to acoustic waves is presented in Fig. 7.15. However, if the plasma is embedded in magnetic fields, as it is in most parts of the solar atmosphere, the framework of MHD may be the appropriate approach. These coronal heating theories are called MHD heating mechanisms.

The ultimate viscous dissipation in MHD models invoke Joule heating or, to a somewhat less extent, viscosity. Examples of energy carriers of magnetic heating are the slow and fast MHD waves, Alfvén waves, magnetoacoustic-gravity waves, current sheets, etc. There is an interesting concept put forward by De Pontieu et al. (2005), where the direct energy coupling and transfer from the solar photosphere into the corona are demonstrated by simulations and TRACE observations; see De Pontieu et al. (2003).

Fig. 7.16 Resonance heating
in corona; from Ulmschneider
(2003)

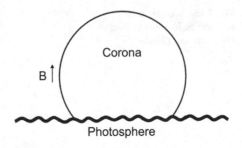

One of the popular alternative MHD heating mechanisms is the selective decay
of a turbulent cascade of magnetic field (Gomez et al. 2000; Hollweg 2002; van
Ballegooijen 1986). Based on the time scales involved, an alternative classification
of the heating mechanism can be constructed. If the characteristic time scale of
the perturbations is less than the characteristic times of the back reaction in a
nonmagnetized plasma, acoustic waves are good approximations describing the
energy propagation; if, however, the plasma is magnetized and perturbation time
scales are small, one uses the alternating current (AC-), which has low frequencies
(hydrodynamic pulses may be appropriate) in a nonmagnetized plasma, while if the
external driving forces (e.g., photospheric motions) operate on longer time scales
compared to dissipation and transit times, very narrow current sheets are built
up, resulting in direct current (DC-) heating mechanisms in magnetized plasmas
(Priest and Forbes 2000). After it was discovered that the coronal plasma is
heavily embedded in magnetic fields, the relevance of the hydrodynamic heating
mechanisms for the corona part of the atmosphere was reevaluated.

Resonance heating occurs when, upon reflection of Alfvén waves at the two foot
points of the coronal loops, one has constructive interference (see Fig. 7.16). For a
given loop length $l_{\|}$ and Alfvén speed v_A, resonance occurs, when the wave period is
$mP = 2l_{\|}/v_A$, m being a positive integer. Waves that fulfill the resonance condition
are trapped and after many reflections are dissipated by Joule, thermal conductive,
or viscous heating.

The current belief is that hydrodynamic heating mechanisms could still contribute to atmospheric heating of the Sun but only at lower layers, that is, possibly in
the chromosphere and up to the magnetic canopy (De Pontieu et al. 2004; De Pontieu
and Erdelyi 2006; Erdelyi 2006). At least as a first approximation, the plasma is
considered frozen-in in the various magnetic structures in the hot solar atmosphere.
The magnetic field plays a central and key role in the dynamics and energetics of
the solar corona. High-resolution satellite observations show the magnetic building
blocks that seem to be in the form of magnetic flux tubes in the solar atmosphere.
These flux tubes expand rapidly in height because of the strong drop in density.
Magnetic fields fill the solar atmosphere almost entirely at about 1,500 km above
the photosphere. A more elaborate discussion on the heating of the chromosphere
and the corona in the Sun is found in Ulmschneider (2003).

7.7 EIT and Moreton Waves

Solar flares and coronal mass ejections (CMEs) are explosive phenomena in the solar atmosphere, capable of launching global large-amplitude coronal disturbances and shock waves. The longest-known signatures of coronal shock waves are radio type II bursts (Payne-Scott et al. 1947; Wild and McCreaky 1950) and Moreton waves (Moreton 1960; Moreton and Ramsey 1960). A type II burst is excited at the local plasma frequency (and/or harmonic) by a fast-mode MHD shock (e.g., Nelson and Melrose 1985, and references therein). As the shock propagates outward through the corona, the emission drifts slowly (in comparison with fast-drift type III emission) toward lower frequencies due to decreasing ambient density. Radial velocities, inferred from the emission drift rates by using various coronal density models, are found to be on the order of $1,000\,\mathrm{km\,s^{-1}}$. The Moreton wave is a large-scale wave-like disturbance of the chromosphere, observed in H_α, which propagates out of the flare site at velocities also in the order of $1,000\,\mathrm{km\,s^{-1}}$. In this respect, it is worth noting that the first indications of global coronal disturbances were provided by flare-associated activations of distant filaments (Dodsch 1949; see also Ramsey and Smith 1966).

The MHD model unifying both phenomena in terms of the fast-mode shock wave was proposed by Uchida (1974). According to Uchida's (1968) sweeping-skirt scenario, the Moreton wave is the surface track of the fast-mode MHD coronal shock propagating out of the source region along valleys of low-Alfvén velocity, namely, being refracted from the high-Alfvén-velocity regions and enhanced in low-velocity regions. At larger heights, the shock causes the type II burst. An EUV counterpart of the Moreton wave was reported by Neupert (1989), and a decade later these coronal disturbances were directly imaged by the Extreme ultraviolet Imaging Telescope (EIT; Delaboudiniere et al. 1995) on the Solar and Heliospheric Observatory (SOHO). The discovery of EIT waves (Moses et al. 1997; Thompson et al. 1998) prompted a search for wave signatures in other spectral domains. Soon, the Moreton wave–associated disturbances were revealed in soft X-rays (Narukage et al. 2007; Khan and Aurass 2002; Hudson et al. 2003; Warmuth et al. 2005), He I $10\,830\,A°$ (Gilbert et al. 2001; Vrasnak et al. 2002; Gilbert and Holzer 2004), and microwaves (Warmuth et al. 2004; White and Thompson 2005); for an overview and historical background.

It is important to note that some of propagating EUV signatures denoted as EIT waves are probably not wave phenomena, but rather a consequence of some other processes related to the large-scale magnetic field reconfiguration. EIT disturbances of this kind are usually much slower and more diffuse than those representing the coronal counterpart of the Moreton waves and those accompanied by type II bursts Such nonwave events either could be a consequence of the CME associated field-line opening or could be caused by various forms of coronal restructuring driven by the eruption.

Generally, large-amplitude MHD waves in the corona are tightly associated with CMEs and flares. The source of the coronal wave seems clear in events where the

CME is accompanied only by a very weak/gradual flare-like energy release. In such cases, the type II bursts characteristically start in the frequency range well below 100 MHz.

Typically, Moreton waves appear as arc-shaped chromospheric disturbances, propagating away from the flare site at speeds on the order of $1,000 \, km \, s^{-1}$. The wavefronts are seen in emission in the center and the blue wing of the H_α line, whereas in the red wing they appear in absorption. This is interpreted as a compression and subsequent relaxation of the chromosphere, due to the increased pressure behind the coronal shock sweeping over the chromosphere (Uchida 1968). Such behavior strongly favors the interpretation in terms of freely propagating large-amplitude simple waves. Further supporting evidence for such an interpretation is found in the deceleration of the wavefront, elongation of the perturbation profile, and decreasing amplitude of the disturbance.

In this respect, it should be emphasized that there are two ways to form a simple-wave shock pattern. The straightforward option is the formation of the shock by a temporary 3D piston effect, which can be caused either by the flare-volume expansion or by the initial lateral expansion of the CME. On the other hand, distant flanks of a bow shock also have simple-wave characteristics. This suggests that flares that occur in the core of the active region and do not have remote extensions toward quiet regions away from sunspots are not likely to cause a Moreton wave.

The Moreton wavefront usually becomes detectable at distances on the order of 100 Mm from the source region, most often becoming clearly recognizable in the range $100 \approx 150$ Mm. Similar to high-frequency type II bursts, the Moreton wave onset is closely associated with the flare-impulsive phase, usually being delayed by a few minutes. Thus, the onset of the type II burst and the Moreton wave appearance are closely linked, implying that the Moreton wave becomes prominent only after the shock has been formed. Such a short time/distance for shock formation requires an extremely impulsive acceleration of the source region. Since the source region expansion has to be accelerated to a velocity on the order of $1,000 \, km \, s^{-1}$ within a minute or so, this requirement favors the flare scenario, since flares typically develop on a shorter time scale than CMEs.

Solar flares and coronal mass ejections (CMEs) may also be treated as explosive processes that are able to generate large-scale wave-like disturbances in the solar atmosphere. Signatures of large-scale wave-like disturbances were first imaged in the hydrogen H_α spectral line and called Moreton waves after Moreton (1960; see also Moreton and Ramsey 1960). Typically, Moreton waves appear as propagating dark and bright fronts in H_α filter grams and Doppler grams, respectively, which can be attributed to a compression and relaxation of the chromospheric plasma. The disturbance propagates with a speed on the order of $1,000 \, km \, s^{-1}$ (e.g., Moreton and Ramsey 1960), which led to the conclusion that such a phenomenon cannot be of chromospheric origin, but is the surface track of a coronal disturbance, compressing the underlying chromosphere (sweeping-skirt hypothesis; see Uchida 1968). Moreton waves are generally observed to be closely associated with the flare-impulsive phase, which often also coincides with the acceleration phase of the associated CME. Moreton waves are observed to propagate perpendicularly to the magnetic field, and the initial magnetosonic Mach numbers are estimated to fall in the range of $M_{ms} \approx 1.4 - -4$, suggesting that they are at least initially shocked fast-mode waves. These results indicate that Moreton waves are a consequence of shocks formed from large-amplitude waves that decay to ordinary fast magnetosonic waves, which is in line with the flare-initiated blast wave scenario. Further evidence for the close association with shocks is the quasi-simultaneous appearance of Moreton waves and radio type II bursts, which are one of the best indicators of coronal shocks.

Wave-like disturbances were for the first time imaged directly in the corona by the EIT instrument aboard the Solar and Heliospheric Observatory (SOHO), thereafter called EIT waves. They were considered to be the coronal manifestation of the Moreton wave, but statistical studies revealed discrepancies in their velocities. EIT waves were found to be two to three times slower than Moreton waves. Today the relationship between EIT waves and Moreton waves and their generation mechanism are very much debated.

This chapter ends with an introduction to the different types of waves that are present in the Sun, with an emphasis on the observations, both from ground-based instruments and from instruments placed on satellites. Again, the choice of material is rather restricted to the authors' understanding and confidence level. The reader is advised to look into the papers cited for further details and a better understanding.

Chapter 8
Helioseismology

The theory of helioseismology, which deals with the study of solar oscillations, has proved to be an extremely successful tool for the investigation of the internal structure and dynamics of the Sun. It has done much to improve our understanding of the interior of the Sun, testing the physical inputs used to model stellar interiors and providing a detailed map of the Sun's structure and internal rotation. This in turn has greatly influenced theories of the solar magnetic dynamo. These studies bridge valuable information between solar physics and that of the structure and evolution of other stars. Recent developments include new local techniques for unprecedented studies of subsurface structures and flows in emerging active regions, under sunspots, and even of active regions on the far side of the Sun. These developments hold the possibility of a real understanding of how the interior is linked to solar magnetic activity in the corona and heliosphere. Studies such as those of the deep solar interior are on the verge of becoming possible for other stars exhibiting similar multimode oscillations. Many figures and equations presented in this chapter is also found in Dalsgard (2002, 2003). An excellent review of the recent results of helioseismology is found in Basu and Antia (2008).

Historically, in 1962, Leighton et al. discovered patches of the surface of the Sun moving up and down, with a velocity of the order of $15 \, \text{cm} \, \text{s}^{-1}$ (in a background noise of $330 \, \text{m} \, \text{s}^{-1}$), with periods near 5 min. Termed the "5-min oscillation," the motions were originally believed to be local in character and related to turbulent convection in the solar atmosphere. A few years later, Ulrich (1970) and, independently, Leibacher and Stein (1971) suggested that the phenomenon is global and that the observed oscillations are the manifestation at the solar surface of resonant sound waves (pressure modes, or *p-modes*) traveling in the solar interior. There are ≈ 10 million resonant *p-modes* in the Sun, with periods ranging from a few minutes to hours. Similar to the sound waves trapped in an organ pipe, the *p*-modes are effectively trapped in a spherical-shell cavity defined by the surface and an inner turning radius that depends on the physical characteristics of the mode itself and the interior. Linear theory may be used to describe the oscillations because their amplitudes are small. Spherical harmonics, characterized by their degree *l* and

A. Satya Narayanan, *An Introduction to Waves and Oscillations in the Sun*, Astronomy and Astrophysics Library, DOI 10.1007/978-1-4614-4400-8_8,
© Springer Science+Business Media New York 2013

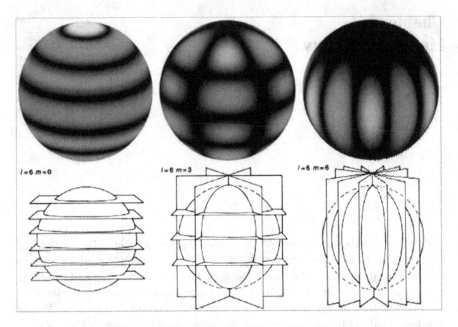

Fig. 8.1 The spherical harmonics in the Sun; from J. Leibacher, NSO, USA

azimuthal order m, describe the angular component of the mode eigenfunctions. The properties of spherical harmonics used to describe nonradial oscillations in the Sun are illustrated in Fig. 8.1, while the relationship among the period, frequency, and wavelength is given in Fig. 8.2. In addition, the radial order n is related to the number of nodes in the radial eigenfunction. In general, low-l-modes penetrate more deeply inside the Sun. They have deeper inner turning radii than higher l-valued *p-modes*. It is this property that gives *p*-modes remarkable diagnostic power for probing layers of different depths in the solar interior.

8.1   Equations of Motion

The basic assumption in helioseismology is that the gas may be treated as a continuum, so that the fluid properties such as density $\rho(\mathbf{r},t)$, pressure $p(\mathbf{r},t)$, and any other thermodynamic quantities that may be required are functions of the position $\mathbf{r}$ and time t. We use the Eulerian description, wherein the fluid motion is seen by a stationary observer. Sometimes it is also useful to use the Lagrangian description, where the observer follows the motion of the fluid. A given element of gas is generally labeled by its initial position $\mathbf{r}_0$ and its motion by $\mathbf{r}(t,\mathbf{r}_0)$. We define its velocity as

$$\mathbf{v}(\mathbf{r},t) = \frac{d\mathbf{r}}{dt}$$

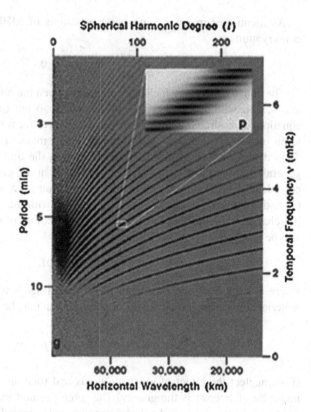

Fig. 8.2 The relationship among the period, frequency, and wavelength; from GONG, USA

for a fixed $\mathbf{r}_0$, which is equivalent to the Euler description. The convective derivative or the material derivative of a quantity ϕ, observed during its motion, is given by

$$\frac{d\phi}{dt} = \left(\frac{\partial\phi}{\partial t}\right)_{\mathbf{r}} + \nabla\phi \cdot \frac{d\mathbf{r}}{dt} = \frac{\partial\phi}{\partial t} + \mathbf{v}\cdot\nabla\phi. \tag{8.1}$$

The different properties such as velocity, density, and pressure are often expressed in terms of vector and scalar fields. We make use of the following vector algebra in dealing with the fields. For example, the Gauss theorem states that

$$\int_{\partial V} \mathbf{a}\cdot\mathbf{n}dA = \int_V \nabla\cdot\mathbf{a}dV, \tag{8.2}$$

where V is the volume, with ∂V the surface, $\mathbf{n}$ is the normal directed outward to ∂V, and $\mathbf{a}$ is any vector field. One can simplify the above equation to yield

$$\int_{\partial V} \phi\mathbf{n}dA = \int_V \nabla\phi dV. \tag{8.3}$$

As mentioned in Chap. 3 on basic equations of MHD, we express the mass conservation as

$$\frac{\partial \rho}{\partial t} + \nabla \cdot (\rho \mathbf{v}) = 0. \tag{8.4}$$

The above equation, which is a balance between the rate of change of a quantity in a volume with its flux density into the volume, is one of the typical conservation equations in hydrodynamics. The other alternative form for the above equation in terms of $\rho = 1/V$, where V is the volume of unit mass, also holds.

In dealing with astrophysical bodies, that is, the Sun and other stars, one can generally ignore the internal friction (viscosity) in the gas. The other forces to be considered are (1) surface forces (the pressure on the surface of the volume) and (2) body forces. For simplicity, we will not deal with the effect of magnetic field in this chapter. With the above assumptions, the equations of motion in hydrodynamics will be written as

$$\rho \frac{d\mathbf{v}}{dt} = -\nabla p + \rho \mathbf{f}, \tag{8.5}$$

where $\mathbf{f}$ is the body force per unit mass, which needs to be specified. Using the material convective derivative, the above equation may be simplified to yield

$$\rho \frac{\partial \mathbf{v}}{\partial t} + \rho \mathbf{v} \cdot \nabla \mathbf{v} = -\nabla p + \rho \mathbf{f}. \tag{8.6}$$

If we neglect the effect of magnetic fields and rotation, the only body force that might be of interest is the gravity. The force per unit mass from gravity may be written in terms of the gradient of a gravitational potential as follows:

$$\mathbf{g} = \nabla \Phi, \tag{8.7}$$

where the potential Φ satisfies the Poisson equation

$$\nabla^2 \Phi = -4\pi G \rho, \tag{8.8}$$

which has a standard integral solution given by

$$\Phi(\mathbf{r}, t) = G \int_V \frac{\rho(\mathbf{r}', t) dV}{|\mathbf{r} - \mathbf{r}'|}. \tag{8.9}$$

The basic equations are completed by making use of the energy equation. For this we will use a relation between the pressure p and the density ρ, in terms of a thermodynamic relation. From the first law of thermodynamics, we have

$$\frac{dq}{dt} = \frac{dE}{dt} + p \frac{dV}{dt}. \tag{8.10}$$

Here, dq/dt is the rate of heat loss or gain, and E the internal energy per unit mass. V, as defined earlier, is the specific volume $V = 1/\rho$. The above equation may be interpreted as that the heat gain is transferred partly into the internal energy and partly into work for expanding or compressing the gas.

By introducing the adiabatic exponents Γ_1, Γ_2, and Γ_3 as given below,

$$\Gamma_1 = \left(\frac{\partial \ln p}{\partial \ln \rho}\right)_{ad}, \quad \frac{\Gamma_2 - 1}{\Gamma_2} = \left(\frac{\partial \ln T}{\partial \ln p}\right)_{ad}, \quad \Gamma_3 - 1 = \left(\frac{\partial \ln T}{\partial \ln \rho}\right)_{ad}, \tag{8.11}$$

the energy equation can be simplified to yield

$$\frac{dq}{dt} = \frac{1}{\rho(\Gamma_3 - 1)}\left(\frac{dp}{dt} - \frac{\Gamma_1 p}{\rho}\frac{d\rho}{dt}\right) \tag{8.12}$$

$$= c_p\left(\frac{dT}{dt} - \frac{\Gamma_2 - 1}{\Gamma_2}\frac{T}{p}\frac{dp}{dt}\right) \tag{8.13}$$

$$= c_V\left[\frac{dT}{dt} - (\Gamma_3 - 1)\frac{T}{\rho}\frac{d\rho}{dt}\right]. \tag{8.14}$$

In most interiors of stars, the approximation of a fully ionized gas is valid. This implies that the adiabatic index takes on the value $5/3$. For the motion that is adiabatic, the energy equation simplifies considerably as follows:

$$\frac{dp}{dt} = \frac{\Gamma_1 p}{\rho}\frac{d\rho}{dt}. \tag{8.15}$$

The above equation, with the equation of mass continuity (8.4), the equations of motion (8.6), and Poisson's equation (8.8), will form a set of equations for adiabatic motions. The theory of helioseismology will deal mostly with the above equations.

8.2 Equilibrium Structure

To begin with, we shall assume that the equilibrium state is static, so that all time derivatives are ignored. Moreover, we shall assume that there are no velocities. The equation of continuity, which is the conservation of mass, is trivially satisfied. The equations of motion reduce to the hydrostatic case:

$$\nabla p_0 = \rho_0 g_0 = \rho_0 \nabla \Phi_0. \tag{8.16}$$

The subscript "0" denotes the equilibrium quantities. The Poisson equation does not change, being a linear equation:

$$\nabla^2 \Phi_0 = -4\pi G \rho_0. \tag{8.17}$$

The energy equation has the following form (without adiabatic approximation):

$$0 = \frac{dq}{dt} = \varepsilon_0 - \frac{1}{\rho_0} \nabla \cdot \mathbf{F}_0. \tag{8.18}$$

Here ε is the rate of energy generation, while $\mathbf{F}$ is the flux of energy.

We shall assume a spherically symmetric equilibrium, so that the structure depends on the radial distance r to the center. We also assume that the gravitational force $\mathbf{g}_0$ is given by $\mathbf{g}_0 = -g_0 \mathbf{a}_r$, where $\mathbf{a}_r$ is the unit vector, radially directed outward. The equation of motion becomes

$$\frac{dp_0}{dr} = -g_0 \rho_0. \tag{8.19}$$

The Poisson equation is integrated to yield

$$g_0 = \frac{G}{r} \int_0^r 4\pi \rho_0 r'^2 dr' = \frac{Gm_0}{r^2}. \tag{8.20}$$

$m_0(r)$ is the mass in the interior of the sphere to r. The flux is directed radially outward, $\mathbf{F} = F_{r,0} \mathbf{a}_r$, and the energy equation is written as

$$\rho_0 \varepsilon_0 = \frac{1}{r^2} \frac{d}{dr}(r^2 F_{r,0}) = \frac{1}{4\pi r^2} \frac{dL_0}{dr},$$

where $L_0 = 4\pi r^2 F_{r,0}$ is the total flow of energy through the sphere with radius r. Combining with the energy equation yields

$$\frac{dL_0}{dr} = 4\pi r^2 \rho_0 \varepsilon_0. \tag{8.21}$$

Equations (8.19)–(8.21) describe the equilibrium for stellar structures.

8.3 Perturbation Analysis

The equilibrium state is perturbed to small fluctuations so that quantities like pressure may be written as

$$p(\mathbf{r},t) = p_0(\mathbf{r}) + p'(\mathbf{r},t), \tag{8.22}$$

where the perturbation $p'(\mathbf{r},t)$ is small and is a function of position and time, while the equilibrium pressure is a function of position only. The basic equations are linearized using these perturbations in such a way that the perturbed quantities do not contain products of the perturbations.

The perturbed continuity equation may be written as

$$\frac{\partial \rho'}{\partial t} + \nabla \cdot (\rho_0 \mathbf{v}) = 0, \qquad (8.23)$$

and the equations of motion become

$$\rho_0 \frac{\partial \mathbf{v}}{\partial t} = -\nabla p' + \rho_0 \mathbf{g}' + \rho' \mathbf{g}_0, \qquad (8.24)$$

where $\mathbf{g}' = \nabla \Phi'$. The perturbed gravitational potential Φ' also satisfies the Poisson equation

$$\nabla^2 \Phi' = -4\pi G \rho', \qquad (8.25)$$

with the solution given by

$$\Phi' = G \int_V \frac{\rho'(\mathbf{r}', t)}{|\mathbf{r} - \mathbf{r}'|} dV. \qquad (8.26)$$

The perturbed energy equation can be simplified to yield

$$\rho \frac{\partial \delta q}{\partial t} = \delta(\rho \varepsilon - \nabla \cdot \mathbf{F}) = (\rho \varepsilon - \nabla \cdot \mathbf{F})'. \qquad (8.27)$$

For adiabatic motions, one can neglect the heating term so that the energy equation simplifies to

$$\frac{\partial \delta p}{\partial t} - \frac{\Gamma_{1,0} p_0}{\rho_0} \frac{\partial \delta \rho}{\partial t} = 0. \qquad (8.28)$$

Integrating the above equation with respect to time, we have

$$\delta p = \frac{\Gamma_{1,0} p_0}{\rho_0} \delta \rho. \qquad (8.29)$$

Expressing in Eulerian form, the above equation may be written as

$$p' + \delta \mathbf{r} \cdot \nabla p_0 = \frac{\Gamma_{1,0} p_0}{\rho_0} (\rho' + \delta \mathbf{r} \cdot \nabla \rho_0). \qquad (8.30)$$

Equations (8.23)–(8.30) are the linearized perturbation equations to be considered for studying the oscillations in the Sun and the stars.

8.4 Acoustic Waves

The spatially homogeneous equilibrium is one of the simplest possible scenarios one can envision to deal with oscillations. In this case, all the derivatives of the equilibrium quantities vanish. This implies that in Eq. (8.16), gravity must be negligible. This situation is rather ideal and may not be realized exactly. However, if the variation in the equilibrium structure is slow compared to the oscillations one wishes to study, then this approximation may be considered reasonable. To start with, we neglect the gravitational potential perturbations, and so Φ' is small. With the assumption of adiabatic approximation, the equations of motion reduce to

$$\rho_0 \frac{\partial^2 \delta \mathbf{r}}{\partial t^2} = -\nabla p'.$$

Taking divergence of the above relation leads to

$$\rho_0 \frac{\partial}{\partial t^2}(\nabla \cdot \delta \mathbf{r}) = -\nabla^2 p'. \tag{8.31}$$

Expressing p' as a function of ρ' from the adiabatic relation and eliminating the term $\nabla \cdot \delta \mathbf{r}$ from the continuity equation result in

$$\frac{\partial^2 \rho'}{\partial t^2} = \frac{\Gamma_{1,0} p_0}{\rho_0} \nabla^2 \rho' = c_{S0}^2 \nabla^2 \rho', \tag{8.32}$$

where

$$c_{S0}^2 = \frac{\Gamma_{1,0} p_0}{\rho_0}$$

has the dimension of square of the velocity.

Equation (8.32) is the wave equation, whose solution can be determined in terms of plane waves as

$$\rho' = a \, \exp[i(\mathbf{k} \cdot \mathbf{r} - \omega t)]. \tag{8.33}$$

Substituting the above plane-wave solution into the wave equation results in

$$-\omega^2 \rho' = c_{S0}^2 \nabla \cdot (i\mathbf{k}\rho') = -c_{S0}^2 |\mathbf{k}|^2 \rho'.$$

The dispersion relation from the above expression can be immediately written as

$$\omega^2 = c_{S0}^2 |\mathbf{k}|^2. \tag{8.34}$$

The waves describing the above relation are the plane sound waves. The adiabatic sound speed c_{S0} is the speed of propagation of the waves. The flow variables, such as the density, pressure, and displacement vector, can be written as

$$\rho' = a \, \cos(\mathbf{k} \cdot \mathbf{r} - \omega t)$$

$$p' = c_{S0}^2 \, a \, \cos(\mathbf{k} \cdot \mathbf{r} - \omega t)$$

$$\delta \mathbf{r} = \frac{c_{S0}^2}{\rho_0 \omega^2} \, a \, \cos(\mathbf{k} \cdot \mathbf{r} - \omega t + \pi/2)\mathbf{k}. \tag{8.35}$$

8.5 Internal Gravity Waves

In this case, we shall study a layer of gas stratified by gravity. This will induce a pressure gradient in the gas. However, we continue to assume that the equilibrium quantities vary slowly so that their gradients can be ignored compared to the gradients of the perturbations. Here also, we neglect the perturbation of the gravitational potential. We seek solutions in the form of plane waves with much lower frequencies compared to the acoustic waves. We assume perturbations of the form

$$\exp[i(\mathbf{k} \cdot \mathbf{r} - \omega t)],$$

with the gradient of equilibrium pressure and density as

$$\nabla p_0 = \frac{dp_0}{dr} \mathbf{a}_r \quad \nabla \rho_0 = \frac{d\rho_0}{dr} \mathbf{a}_r. \tag{8.36}$$

In addition to the above, we separate the wave vector $\mathbf{k}$ and the displacement $\delta \mathbf{r}$ into radial and tangential components as

$$\delta \mathbf{r} = \xi_r \mathbf{a}_r + \xi_t$$

$$\mathbf{k} = k_r \mathbf{a}_r + \mathbf{k}_t. \tag{8.37}$$

The radial and tangential components of the equations of motion are

$$-\rho_0 \omega^2 \xi_r = -ik_r p' - \rho' g_0 \tag{8.38}$$

$$-\rho_0 \omega^2 \xi_t = -i\mathbf{k}_t p', \tag{8.39}$$

with the continuity equation, which can be written as

$$\rho' + \rho_0 ik_r \xi_r + \rho_0 i\mathbf{k}_t \cdot \xi_t = 0. \tag{8.40}$$

Using Eqs. (8.39) and (8.40), the pressure perturbation can be written as

$$p' = \frac{\omega^2}{k_t^2} (\rho' + ik_r \rho \xi_r). \tag{8.41}$$

This may be used to derive an equation for the radial component of the displacement as

$$-\rho_0 \omega^2 \xi_r = -i\frac{k_r}{k_t^2} \omega^2 \rho' + \omega^2 \rho_0 \frac{k_r^2}{k_t^2} \xi_r - \rho' g_0. \tag{8.42}$$

Since we are interested in small-frequency perturbations, the first term in ρ' may be ignored compared to the second term in the above equation, which leads to

$$\rho_0 \omega^2 \left(1 + \frac{k_r^2}{k_t^2} \right) \xi_r = \rho' g_0. \tag{8.43}$$

The physical meaning of the above expression is that the buoyancy acting on the density perturbation provides a vertical force $\rho' g_0$ per unit volume, which drives the motion. The left-hand side, which deals with the vertical acceleration times the mass ρ_0 per unit volume, is modified by the wavenumbers. The adiabatic relationship on simplification yields

$$\rho' = c_{S0}^{-2} p' + \rho_0 \delta \mathbf{r} \cdot \left(\frac{1}{p_0 \Gamma_{1,0}} \nabla p_0 - \frac{1}{\rho_0} \nabla \rho_0 \right). \tag{8.44}$$

Inserting the expression for ρ' from Eq. (8.44) into Eq. (8.43) yields

$$\omega^2 \left(1 + \frac{k_r^2}{k_t^2} \right) \xi_r = N^2 \xi_r, \tag{8.45}$$

where

$$N^2 = g_0 \left(\frac{1}{\Gamma_{1,0}} \frac{d\ln p_0}{dr} - \frac{d\ln \rho_0}{dr} \right) \tag{8.46}$$

is the square of the buoyancy or Brunt–Vaisala frequency N. The dispersion relation can easily be obtained as

$$\omega^2 = \frac{N^2}{1 + k_r^2/k_t^2}. \tag{8.47}$$

The motion is oscillatory when $N^2 > 0$. The frequency N can be viewed as that obtained in the limit of infinite k_t, namely, for infinitely small tangential wavelength. One can also interpret the condition $N^2 > 0$ as follows:

$$\frac{d \ln \rho_0}{d \ln p_0} > \frac{1}{\Gamma_{1,0}}. \tag{8.48}$$

If it is not satisfied, then ω is imaginary, so that the motion grows exponentially with time. This is usually interpreted as being due to linear convective instability. The motion in general grows until it breaks down into turbulence due to nonlinear effects. It is important to realize that gravity waves do not generally propagate into the convective regions. Thus, their detection is far from simple.

In addition to the internal gravity waves, one can in principle discuss the type of gravity waves caused due to a sharp density discontinuity. These waves are in general referred to as the surface gravity waves. In order to derive the dispersion relation for the surface gravity waves, we consider a liquid of constant density ρ_0 with a free surface. The pressure on the surface is assumed to be constant for simplicity. It is assumed that the layer is infinitely deep in extent. Also, it is assumed that the fluid is incompressible and that the density perturbation $\rho' = 0$. The equation of continuity reduces to

$$\nabla \cdot \mathbf{v} = 0. \tag{8.49}$$

The gravitational force is assumed to be uniform and directed vertically downward to enable buoyancy for the fluid. As in the previous case, we shall ignore the perturbation to the gravitational potential. The equations of motion can be written as

$$\rho_0 \frac{\partial \mathbf{v}}{\partial t} = -\nabla p'. \tag{8.50}$$

Taking divergence of the above equation yields

$$\nabla^2 p' = 0. \tag{8.51}$$

This is the Laplace equation. Assuming perturbations of the form

$$p'(x, z, t) = f(z)\cos(kx - \omega t), \tag{8.52}$$

where $f(z)$ may be interpreted as the amplitude, which is a function of the vertical coordinate, one can derive the equation governing the amplitude as

$$\frac{d^2 f}{dz^2} = k^2 f. \tag{8.53}$$

The solution of the above equation, assuming it is infinitely deep, yields $f(z) = a \exp(-kz)$. Applying the boundary condition at the free surface and using the relationship between the pressure perturbation and the displacement, we get

$$0 = \left(1 - \frac{g_0 k}{\omega^2}\right) p', \tag{8.54}$$

which implies that the dispersion relation for the surface waves is

$$\omega^2 = g_0 k. \tag{8.55}$$

An interesting observation is that the frequencies of the surface gravity waves depend mainly on the wavelength and gravity, and not on the internal structure of the fluid, in particular its density.

8.6 Equations of Linear Stellar Oscillations

In this section, the equations governing small oscillations around a spherical equilibrium state are presented. Here we explicitly use the spherical symmetry and give the equations for nonradial oscillations. The special case of radial oscillations will also be given.

We shall assume the dependent variables, such as the displacement, pressure perturbation, and gravitational potential, as follows:

$$\xi_r(r,\theta,\phi,t) = \sqrt{4\pi}\tilde{\xi}_r(r)Y_l^m(\theta,\phi)\exp(-i\omega t) \tag{8.56}$$

$$p'(r,\theta,\phi,t) = \sqrt{4\pi}\tilde{p}'(r)Y_l^m(\theta,\phi)\exp(-i\omega t) \tag{8.57}$$

and a similar form for the gravitational potential. Inserting the above into the equations of motion (tangential and radial components), equation of continuity, Poisson's equation, and the energy equation, after simplification, can be written as ordinary differential equations for the amplitude functions as follows:

$$\omega^2\left[\tilde{p}' + \frac{1}{r^2}\frac{d}{dr}(r^2\rho_0\tilde{\xi}_r)\right] = \frac{l(l+1)}{r^2}(\tilde{p}' - \rho_0\tilde{\Phi}') \tag{8.58}$$

$$-\omega^2\rho_0\tilde{\xi}_r' = -\frac{d\tilde{p}'}{dr} - \tilde{p}'g_0 + \rho_0\frac{d\tilde{\Phi}'}{dr} \tag{8.59}$$

$$\frac{1}{r^2}\frac{d}{dr}\left(r^2\frac{d\tilde{\Phi}'}{dr}\right) - \frac{l(l+1)}{r^2}\tilde{\Phi}' = -4\pi G\tilde{\rho}' \tag{8.60}$$

and the energy equation as

$$\left(\delta\tilde{p} - \frac{\Gamma_{1,0}p_0}{\rho_0}\delta\tilde{\rho}\right) = \rho_0(\Gamma_{3,0} - 1)\delta\tilde{q}. \tag{8.61}$$

It is interesting to note that the above equations are independent of the azimuthal order m. This was possible due to the assumption of spherical symmetry. The above set of coupled equations is far from easy to solve. Thus, we shall resort to equations that describe simple linear, adiabatic oscillations. For convenience, the notations of subscript for equilibrium variable and tilde for the amplitudes will be dropped. The resulting equations may be written as

$$\rho' = \frac{\rho}{\Gamma_1 p}p' + \rho\xi_r\left[\frac{1}{\Gamma_1 p}\frac{dp}{dr} - \frac{1}{\rho}\frac{d\rho}{dr}\right]. \tag{8.62}$$

The equation for the displacement can be written by eliminating the density perturbation, so that we have

$$\frac{d\xi_r}{dr} = -\left(\frac{2}{r} + \frac{1}{\Gamma_1 p}\frac{dp}{dr}\right)\xi_r + \frac{1}{\rho}\left[\frac{l(l+1)}{\omega^2 r^2} - \frac{1}{c_S^2}\right]p' - \frac{l(l+1)}{\omega^2 r^2}\Phi'. \tag{8.63}$$

The square of the adiabatic sound speed $c_S^2 = \Gamma_1 p/\rho$. Introducing the characteristic acoustic frequency as S_l, we have

$$S_l^2 = \frac{l(l+1)c_S^2}{r^2} = k_t^2 c_S^2. \tag{8.64}$$

The equations for the pressure perturbation and the gravitational potential can be written as follows:

$$\frac{dp'}{dr} = \rho(\omega^2 - N^2)\xi_r + \frac{1}{\Gamma_1 p}\frac{dp}{dr}p' + \rho\frac{d\Phi'}{dr} \tag{8.65}$$

and

$$\frac{1}{r^2}\frac{d}{dr}\left(r^2\frac{d\Phi'}{dr}\right) = -4\pi G\left(\frac{p'}{c_S^2} + \frac{\rho\xi_r}{g}N^2\right) + \frac{l(l+1)}{r^2}\Phi'. \tag{8.66}$$

The equations for the displacement, pressure perturbation, and gravitational potential given above constitute a complete set of differential equations. For the case of purely radial oscillations, the gravitational potential can be eliminated to yield a second-order system in displacement and pressure perturbation.

The boundary conditions for the displacement, the gravitational potential, and the pressure perturbation may be written as follows: For $l > 0$,

$$\xi_r \approx l\xi_t \quad \text{for} \quad r \to 0. \tag{8.67}$$

The surface condition for the continuity of Φ' at the surface $r = R$ may be written as

$$\frac{d\Phi'}{dr} + \frac{l+1}{r}\Phi' = 0 \quad \text{at} \quad r = R. \tag{8.68}$$

The other condition for the pressure perturbation is given by

$$\delta p = p' + \xi_r\frac{dp}{dr} = 0 \quad \text{at} \quad r = R. \tag{8.69}$$

8.7 Properties of Solar Oscillations (Internal)

The equations of motion mentioned in the previous section are solved numerically with the relevant boundary conditions. The discussion and methodology of the numerical procedure are beyond the scope of this book, and the interested reader is urged to look into the literature Dalsgard (2003). However, suffice to say that the nontrivial solutions of the equations with the boundary conditions will exist only for specific values of the frequency ω, which in turn will lead to an eigenvalue problem. Each eigenfrequency will denote a mode of oscillation. The corresponding eigenfunctions will represent the variation of the perturbations ξ_r, p', etc., as a function of the radial distance r. The eigenfunctions also relate the observable surface amplitude to the amplitude in the interior of the star.

Most of the modes observed in the Sun are essentially acoustic modes, often of relatively high radial order. In this case, an asymptotic description can be obtained very simply, by approximating the modes locally by plane sound waves, satisfying the dispersion relation

$$\omega^2 = c_S^2 |\mathbf{k}|^2,$$ (8.70)

where $\mathbf{k} = k_r \mathbf{a}_r + \mathbf{k}_t$ is the wave vector. The properties of these modes are solely dependent on the variation of the sound speed as a function of the radial distance. The radial variation of the modes can be described by the following equation:

$$k_r^2 = \frac{\omega^2}{c_S^2} - \frac{L^2}{r^2} = \frac{\omega^2}{c_S^2}\left(1 - \frac{S_l^2}{\omega^2}\right).$$ (8.71)

The normal modes observed as global oscillations on the stellar surface arise through interference between propagating waves. In particular, they share with the waves the total internal reflection at $r = r_t$. It is clear that the closer the lower turning point is located to the center, the lower the degree is or the higher the frequency is. Radial modes, with $l = 0$, penetrate the center, whereas the modes of highest degree observed in the Sun, with $l > 1,000$, are trapped in the outer small fraction of a percent of the solar radius. Thus, the oscillation frequencies of different modes reflect very different parts of the Sun; it is largely this variation in sensitivity that allows the detailed inversion for the properties of the solar interior as a function of the position.

The adiabatic oscillation equations presented in the previous section obviously depend on the basic structure of the equilibrium model. This is more so in the case of the Sun. The coefficients would depend on the equilibrium variables such as ρ, the density; p, the pressure; Γ_1, the adiabatic index; and g, the force due to gravity. For example, if the density $\rho(r)$ as a function of r is known, then one can use Poisson's equation to determine $g(r)$. Once the density and gravity are known, then one can use the equation of hydrostatics to determine $p(r)$. An important point to note is that the determination of the adiabatic index Γ_1 is far from trivial. The thermodynamic state of the chemical composition must be known, and in general, we are not allowed to assume a form for it. A simple rule of thumb is that it must be close to the value of 5/3. Figure 8.3 presents the cyclic frequencies $\nu = \omega/2\text{pi}$ as functions of degree l computed for a normal solar model for the p, g, f models. Details about the normal solar model my be found in (ref). It is evident from Fig. 8.3 that there are two distinct, but slightly overlapping groups of modes, with different frequency behavior for different values of l. For $n > 0$, the upper set of modes appears. These modes are referred to as the p-modes, the pressure modes. The mode $n = 0$ has a similar behavior as the p-modes. However, it is distinct in the sense that for reasonably large values of l, say, greater than 20, its behavior is that of a surface gravity wave of the f-mode. Modes for which $n < 0$ are the g-modes. The g-mode spectrum extends to zero frequency for all degrees, although they get crowded as the value of the degree increases.

Fig. 8.3 Cyclic frequencies ν as functions of degree l; from Dalsgard (2003)

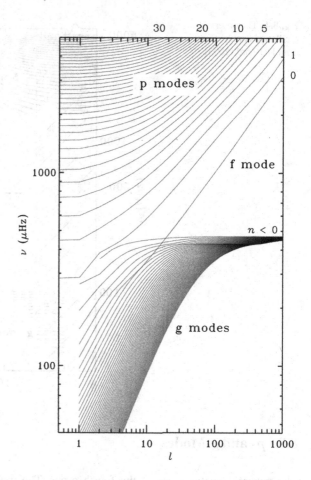

The observed oscillations on the Sun are presented in Fig. 8.4. The 5min oscillation is predominantly seen for larger values of frequencies ν and several degrees of l. Figures 8.5 and 8.6 depict the typical eigenfunctions for g- and p-modes. The quantity plotted in the vertical axis $r\rho^{1/2}\xi_r$ is a measure of the contribution of the energy density from the radial component of velocity. In the case of the p-modes, it is clear that the energy, at least for the low degrees, is present throughout the Sun. Unlike the p-modes (see the propagation diagram, Fig. 8.7), the g-modes are more confined to the interior of the Sun. This makes it difficult to observe easily. Also, its maximum energy is close to the center of the Sun with little change in the overall distribution of energy.

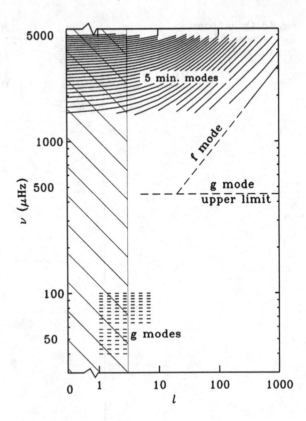

Fig. 8.4 The observed oscillations on the Sun; from Dalsgard (2003)

8.8 *p*- and *g*-Modes

The general equations are of the fourth order. The modes observed in the Sun are either of high radial order or high degree. In such cases it is often possible, in approximate analysis, to make the so-called Cowling approximation, where the perturbation Φ' to the gravitational potential is neglected (Cowling 1941). This can be justified, at least partly, by noting that for modes of high order or high degree, and hence varying rapidly as a function of position, the contributions from regions where ρ' have opposite sign largely cancel in Φ'. In this approximation, the order of the equations is reduced to 2, making them amenable to standard asymptotic techniques (e.g., Ledoux 1962; Vandakurov 1967; Smeyer 1968). A convenient formulation has been derived by Gough (see Deubner and Gough 1984); Gough (1993) in terms of the quantity

$$\Psi = c_S^2 \rho^{1/2} \nabla \cdot \delta \mathbf{r}. \tag{8.72}$$

With the above change in the variable, the oscillation equation can be approximated as

$$\frac{d^2\Psi}{dr^2} = -K(r)\Psi, \tag{8.73}$$

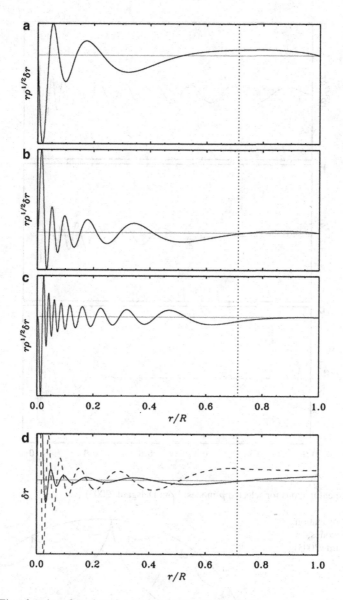

Fig. 8.5 Eigenfunctions for selected g-modes; from Dalsgard (2003)

where

$$K(r) = \frac{\omega^2}{c_S^2}\left[1 - \frac{\omega_c^2}{\omega^2} - \frac{S_l^2}{\omega^2}\left(1 - \frac{N^2}{\omega^2}\right)\right].$$

(8.74)

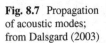

Fig. 8.6 Eigenfunctions for selected p-modes; from Dalsgard (2003)

Fig. 8.7 Propagation
of acoustic modes;
from Dalsgard (2003)

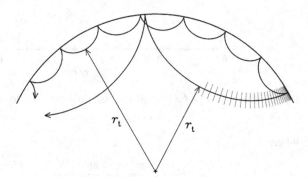

The acoustic cutoff frequency ω_c is given by

$$\omega_c^2 = \frac{c_S^2}{4H^2}\left(1 - 2\frac{dH}{dr}\right),$$ (8.75)

where H is the density scale height.

Using the standard JWKB analysis in the above equations, the modes can be shown to satisfy the following relation:

$$\omega\int_{r1}^{r2}\left[1 - \frac{\omega_c^2}{\omega^2} - \frac{S_l^2}{\omega^2}\left(1 - \frac{N^2}{\omega^2}\right)\right]^{1/2}\frac{dr}{c_S} \approx \pi(n - 1/2),$$ (8.76)

where $r1$ and $r2$ are the adjacent zeros of K such that $K > 0$ between them.

For the *p*-modes, we may conveniently ignore the term in N, except near the surface, the term in ω_c. On the other hand, near the surface, the Lamb frequency $S_l \ll \omega$ for small or moderate l and may be neglected. Assuming $|N^2/\omega^2| \ll 1$, the modes satisfy

$$\omega\int_{r1}^{r2}\left(1 - \frac{\omega_c^2}{\omega^2} - \frac{S_l^2}{\omega^2}\right)^{1/2}\frac{dr}{c_S} \approx \pi(n - 1/2),$$ (8.77)

where $r1 \approx r_t$ and $r2 = R_t$.

In addition to *p*-modes, the observations of solar oscillations also show *f*-modes of moderate and high degree. These modes are approximately divergence-free, with frequencies given by

$$\omega^2 \approx g_s k_h = \frac{GM}{R^3}L,$$ (8.78)

where g_s is the surface gravity. It may be shown that the displacement eigenfunction is approximately exponential, $\xi_r \propto \exp(k_h r)$, as is the case for surface gravity waves in deep water. According to Eq. (8.78), the frequencies of these modes are independent of the internal structure of the star; this allows the modes to be uniquely identified in the observed spectra, regardless of possible model uncertainties. A more careful analysis must take into account the fact that gravity varies through the region over which the mode has substantial amplitude; this results in a weak dependence of the frequencies on the density structure (Gough 1993).

The *g*-modes are trapped in the radiative interior and behave exponentially in the convection zone. In fact, they have their largest amplitude close to the solar center and hence are potentially very interesting as probes of conditions in the deep solar interior. High-degree *g*-modes are very effectively trapped by the exponential decay in the convection zone and are therefore unlikely to be visible at the surface. However, for low-degree modes, the trapping is relatively inefficient, and hence the modes might be expected to be observable if they were excited to reasonable amplitudes. The behavior of the oscillation frequencies can be obtained from Eq. (8.76). In the limit where $\omega \ll N$ in much of the radiative interior, this shows that the modes are uniformly spaced in the oscillation period, with a period spacing that depends on degree n.

In the previous sections, we briefly described the different oscillations that have been dealt with theoretically. In what follows, we briefly give some of the results obtained using helioseismology on the internal structure of the Sun.

Before we get on to the results, let's briefly mention the contributions made by researchers for a better understanding of the internal structure. Duvall and Harvey (1984) studied the intermediate-degree modes, including the rotational splittings. Libbrecht et al. (1990) published a table of solar oscillation frequencies from observations at Big Bear Solar Observatory (BBSO) extending over a few months. This formed the basis of most early studies in helioseismology until 1996, when GONG data became available. Historically, stellar pulsation was first discovered by Fabricius in 1596 when he found periodic variation in the brightness of the star Mira (o Ceti).

The velocity field (line-of-sight component) at the solar surface is typically a few meters per second. The Earth's rotation and orbital velocity are given by 500 m/s, while the rotation of the Sun is 2 km/s. Convective motions are typically 1 km/s, whereas solar oscillations are, in general, <1 m/s. A schematic view of the nature of the oscillations in the Sun is presented in Fig. 8.8.

In order to study the solar oscillations, it was assumed that it is spherically symmetric in nature. Spherical symmetry may be assumed mathematically in the following form:

$$\delta f_{n,l,m}(r,\theta,\phi,t) = K_{n,l,m}(r)Y_l^m(\theta,\phi)e^{i\omega_{n,l,m}t},$$

where l is the degree, m is the azimuthal order, $|m| \leq l$, and n is the radial order.

The horizontal wavenumber is given by $k_h = \sqrt{l(l+1)}/r$ and the frequency by $v = \omega/2\pi = 1/P$.

Observations of the solar oscillations may be categorized into two classes: (1) in integrated light ($l \leq 4$); (2) with spatial resolution ($0 < l \leq N$), where $N =$ no. of pixels. For each of these, one can observe the oscillations either in the line-of-sight velocity via Doppler shift or in intensity. Observations are taken at an interval of 1 min, which gives a Nyquist frequency ($1/(2\triangle t)$) of 8.3333 mHz. The resolution of temporal spectrum is determined by the duration of observations; for example, 1 day gives a resolution of $1/86400 \approx 11.6\,\mu$Hz.

In order to study the internal structure of the Sun, a Global Oscillation Network Group was formed where the instruments were placed at a network of sites on the Earth. The other networks that carried observations are the Birmingham Solar Oscillation Network (BiSON), International Research on Interior of Sun (IRIS), and the Taiwan Oscillation Network (TON). Instruments were also placed in space: Solar and Heliospheric Observatory (SOHO), Global Oscillations at Low Frequencies (GOLF), and Michelson Doppler Imager (MDI).

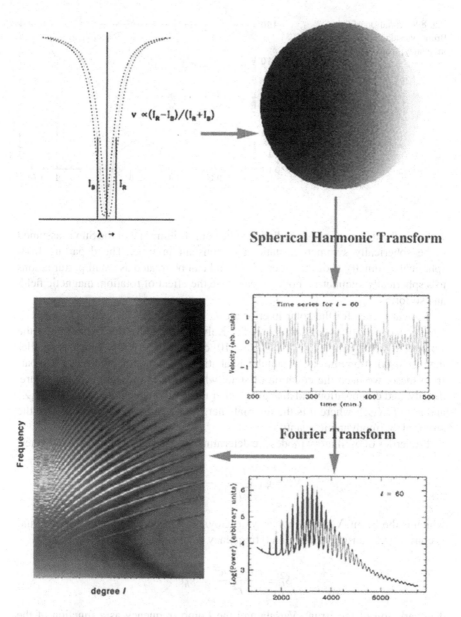

Fig. 8.8 A schematic view of the nature of oscillations in the Sun; from Antia (2007)

Initially, 256×256 CCD was used for observations, and frequencies of 400,000 modes were determined, for $l \leq 150$ and $1 \leq v \leq 4$ mHz. The network operation started on May 7, 1995, and continues even today. Almost 11 years of data covering most of solar cycle 23 have been analyzed. The camera was upgraded to 1024×1024 CCD in 2001.

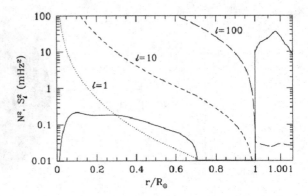

Fig. 8.9 Variation of the Brunt–Vaisala and Lamb frequency; from Antia (2007)

Using the theory of solar oscillations, the equilibrium state of the Sun is assumed to be spherically symmetric, static, and constant in time. The departure from spherical symmetry is of the order of 10^5 and can be treated as small perturbations to a spherically symmetric model to calculate the effect of rotation, magnetic field, and so forth.

Typical values for the solar interior are $L = 10^{10}$ cm, $\rho = 1\,\text{g cm}^3$, $T = 10^6$ K, which give $\tau = 10^7$ years, which is the Kelvin–Helmholtz time scale. Near the surface, $\tau = 10^6\,\text{g cm}^3$, $T = 10^4$ K, $L = 10^7$ cm, and $\tau = 100$ s, which is smaller than the oscillation time scale, and the adiabatic approximation breaks down. Apart from these, we need the equation of state, which relates pressure to temperature, density, and composition and also gives other thermodynamic indices, $p(\rho, T, X, Z)$ and $u(\rho, T, X, Z)$, where u is the internal energy. Derivatives of these will give the required thermodynamic quantities.

Properties of oscillatory modes are determined by two characteristic frequencies:

$$N^2 = g \left(\frac{1}{\Gamma_1 p} \frac{dp}{dr} - \frac{1}{\rho} \frac{d\rho}{dr} \right),$$

which is the Brunt–Vaisala frequency or buoyancy frequency. For $N^2 < 0$, the fluid is convectively unstable. The Lamb frequency is given by

$$S_l^2 = \frac{l(l+1)c^2}{r^2} = k_h^2 c^2.$$

The variation of the Brunt–Vaisala and the Lamb frequency as a function of the nondimensional radius of the Sun is presented in Fig. 8.9.

The variation of the frequency splittings as a function of the frequency in micro-Hertz is presented in Fig. 8.10, while Fig. 8.11 depicts the variation of the density and the sound speed as a function of the solar radii.

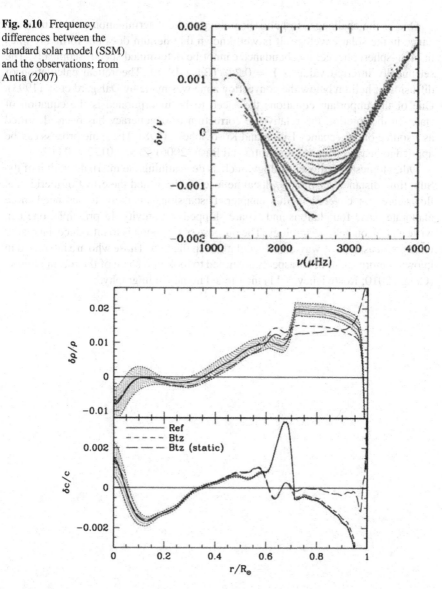

Fig. 8.10 Frequency differences between the standard solar model (SSM) and the observations; from Antia (2007)

Fig. 8.11 Variation of the density and the sound speed as a function of the solar radii; from Antia (2007)

The main goal of helioseismic inferences is obtained by two techniques: (a) the forward technique: compare frequencies of different models with observed frequencies; (b) the inverse technique: using the observed frequencies, infer the internal structure or dynamics.

One of the challenges in helioseismology is the determination of helium abundance in the solar envelope. It is well known that helium does not form any lines in photosphere; hence, its abundance cannot be determined spectroscopically. The seismically inferred value is $Y = 0.249$ (Basu 1998). Theoretical calculation of diffusion of helium below the convection zone was made by Dalsgard et al. (1993). One of the important equations that need to be incorporated is the equation of state. In the interior, the relativistic correction to degeneracy has been identified as a source of discrepancy Elliott and Kosovichev (1998). The same process can be applied to heavy elements, too Antia and Basu (2006): $Z = 0.0172 \pm 0.002$.

Other topics on helioseismology, such as determining the internal rotation of the Sun, time-distance seismology, local helioseismology, and the effect of meridional flows, are not covered in this chapter. Discussions on these topics need more elaborate basic foundations and so are skipped for brevity. In principle, one can write a book on helioseismology. The purpose of this book is to introduce the reader to different aspects of waves and oscillations in the Sun. Those who are interested in knowing more about other aspects are urged to look into some of the recent reviews (Gough 2010; Kosovichev 2011) mentioned in the Bibliography.

References

Abdelatif, T.E.: Ph.D. Thesis, University of Rochester (1985)
Ali, S., Ahmed, Z., Mirza, A.M., Ahmed, I.: Phys. Lett. A **373**, 2940 (2009)
Aller, L.H.: Atoms, Stars and Nebulae, 3rd edn. Cambridge University Press, Cambridge (1998)
Ando, H., Osaki, Y.: PASJ **27**, 581 (1975)
Antia, H.M., Basu, S.: Astrophys. J. **644**, 1292 (2006)
Antia, H.M.: Lectures at Kodaikanal, India (2007)
Aschwanden, M.J.: Sol. Phys. **111**, 113 (1987)
Aschwanden, M.J., Benz, A.O., Montello, M.L.: ApJ **431**, 432 (1994)
Aschwanden, M.J., Fletcher, L., Schrijver, C.J., Alexander, D.: ApJ **520**, 880 (1999)
Aschwanden, M.J., Brown, J.C., Kontar, E.P.: Sol. Phys. **210**, 389 (2002)
Aschwanden, M.J., Schrijver, C.J., De Pontieu, B., Title, A.M.: Sol. Phys. **206**, 99 (2002a)
Aschwanden, M.J.: Astro. Phys. **9505 A** (2003)
Aschwanden, M.J.: ESA SP **575**, 97 (2004)
Aschwanden, M.J., Lemen, J., Nitta, N., Metcalf, T., Wueslar, J., Alexander, D.: Am. Geophys. Union (2004)
Aurass, H., Klein, K.L., Ya Zlotnik, E., Zaitsev, V.V.: A & A **410**, 1001 (2003)
Ballai, I., Ruderman, M.S., Erdelyi, R.: Phys. Plasmas **5**, 252 (1998)
Ballai, I., Erdelyi, R., Pinter, B.: ApJ **633**, L 145 (2005)
Basu, S.: Month. Not. Roy, Astron. Soc. **298**, 719 (1998)
Beckers, J.M., Tallant, P.E.: Sol. Phys. **7**, 351 (1969)
Beckers, J.M., Schwarz, R.B.: Sol. Phys. **27**, 61 (1972)
Bennett, K., Roberts, B., Narain, U.: Sol. Phys. **185**, 41 (1999)
Banerjee, D., Erdelyi, R., Oliver, R., O Shea, E.: Sol. Phys. **246**, 3 (2007)
Banerjee, D., Perez-Suarez, D., Doyle, J.G.: A & A **501**, L15 (2009)
Basu, S., Antia, H.M.: Phys. Rep. **1** (2008)
Beckers, J.M., Tallant, P.E.: Sol. Phys. **7**, 351 (1969)
Bellan, P.M.: Fundamentals of Plasma Physics. Cambridge University Press, Cambridge (2006)
Bhatnagar, A., Livingston, W.C., Harvey, A.W.: Sol. Phys. **27**, 80 (1972)
Bhatnagar, A., Tanaka, K.: Sol. Phys. **24**, 87 (1972)
Bocchialini, K., Baudin, F.: A & A **299**, 893 (1995)
Bogdan, T.J., Carlson, M., Hansteen, V.H., McMurry, A., Rosenthal, C.S., Johnson, M., Petty-Powell, S., Zita, E.J., Stein, R.F., McIntosh, S.W., Nordlund, A.: ApJ **599**, 626 (2003)
Bogdan, T.J., Judge, P.G.: Roy. Soc. Lond. Trans. **364**, 313 (2006)
Bray, R.J., Loughhead, R.E., Durrant, C.J.: The Solar Granulation, 2nd edn. Cambridge University Press, Cambridge (1984)
Browining, P.K., Priest, E.R.: A & A **131**, 283 (1984)

A. Satya Narayanan, *An Introduction to Waves and Oscillations in the Sun*, Astronomy and Astrophysics Library, DOI 10.1007/978-1-4614-4400-8,
© Springer Science+Business Media New York 2013

Bruckner, G.E., et al.: Sol. Phys. **162**, 357 (1995)

Cally, P.S.: Sol. Phys. **88**, 77 (1983)

Cargil, P.J.: Sol. Phys. **147**, 263 (1993)

Carlsson, M., Judge, P.G., Wilhelm, K.: ApJ **486**, L63 (1997)

Carlsson, M., Stein, R.F.: ApJ **481**, 500 (1997)

Carlsson, M., Stein, R.F.: ApJ **481**, 500 (1997)

Carter, B.K., Erdelyi, R.: A & A **475**, 323 (2007)

Carter, B.K., Erdelyi, R.: A & A **481**, 239 (2008)

Chandrasekhar, S., Fermi, E.: ApJ **118**, 116 (1953)

Chandrasekhar, S.: ApJ **124**, 232 (1956)

Chandrasekhar, S., Kendall, P.C.: ApJ **126**, 457 (1957)

Chandrasekhar, S.: Hydrodynamic and Hydromagnetic Stability. Oxford University Press, Oxford (1961)

Chang, P., Quataert, E.: Month. Not. R. Astron. Soc. **403**, 246 (2010)

Chaplin, W.J.: Music of the Sun (Helioseismology). One World Publications (2006)

Chen, F.F.: Introduction to Plasma Physics. Plenum, NY (1977)

Clark, S.: The Sun Kings. Princeton University Press, Princeton (2007)

Claverio, A., Isaak, G.R., McLeod, C.P., van der Raay, H.B., Cortes, T.R.: Nature **282**, 591 (1979)

Cram, L.E., Thomas, J.H.: The Physics of Sunspots. Sacramento Peak Observatory, Sunspot (1981)

Crooker, N., Joselyn, J.A., Feynman, J.: Coronal Mass Ejections. American Geophysical Union, Washington, DC (1997)

Csik, A.J., Cadez, V.M., Goossens, M.: A & A **339**, 215 (1998)

Cowling, T.G.: MNRAS, **101**, 367 (1941)

Dalsgard, J.C., Proffitt, C.R., Thompson, M.J.: Astrophys. J. **403**, L75 (1993)

Dalsgard, J.C.: Lecture Notes on Stellar Oscillations, (2003)

Dalsgard, J.C.: Rev. Modern Phys. **1** (2002)

Dame, L., Gouttebroze, P., Malherbe, J.M.: A & A **130**, 331 (1984)

David Rathinavelu, G., Sivaraman, G., Narayanan, A.S.: Plasma Phys. Rep. **35**, 976 (2009)

David Rathinavelu, G., Sivaraman, G., Narayanan, A.S.: Astrophysics Space Science Proceedings, vol. 510. Springer, Berlin (2010)

David Rathinavelu, G., Sivaraman, G., Narayanan, A.S.: J. Phys. Conf. Ser. **208**, 012071 (2010a)

Delaboudiniere, J.P., et al.: Sol. Phys. **162**, 291 (1995)

Zanna, L.D., Velli, M., Londrillo, P.: A & A **367**, 705 (2001)

Dendy, R.O.: Plasma Dynamics. Oxford University Press, Oxford (1990)

De Forest, C.E., Gurman, J.B.: ApJ **501**, L 217 (1998)

De Moortel, I.: Roy. Soc. Lond. Trans. **363**, 2743 (2005)

De Pontieu, B., Erdelyi, R., de Wijn, A.G.: ApJ **595**, L 63 (2003)

De Pontieu, B., Erdelyi, R., De Moortel, I.: ApJ **624**, L 61 (2005)

De Pontieu, B., Erdelyi, R., De Moortel, I., Metcalf, T.: AGU Abstract SH13A - 1142 American Gephysical Union preprint (2004)

De Pontieu, B., Erdelyi, R.: Roy. Soc. Lond. Trans. **364**, 383 (2006)

Deubner, F.L.: IAU Symposium No. 56, held at Australia (1973)

Deubner, F.L.: A & A **44**, 371 (1975)

Deubner, F.L., Gough, D.: Ann. Review. Astron. Astrophys. **22**, 593 (1984)

Diaz, A.J., Ballester, J.L.: A & A **402**, 781 (2003)

Dodsch, H.W.: ApJ **110**, 382 (1949)

Durrant, C.J.: The Atmosphere of the Sun. Adam Hilger, Bristol (1988)

Duvall, T.L., Harvey, J.W.: Nature **302**, 42 (1983)

Duvall, T.L., Harvey, J.W.: Nature, **310**, 19 (1984)

Edwin, P.M., Roberts, B.: Sol. Phys. **88**, 179 (1983)

Elliott, J.R., Kosovichev, A.G.: Astrophys. J. **500**, L199 (1998)

Erdelyi, R., Doyle, J.G., Perez, M.E., Wilhelm, K.: A & A **337**, 287 (1998)

Erdelyi, R., Ballai, I.: Sol. Phys. **186**, 67 (1999)

Erdelyi, R., Varga, E., Zetenyi, M.: 9th ESPM, Florence, Italy, 269 (1999)

Erdelyi, R.: ESA, ASP **624**, 72 (2006)

Erdelyi, R., Carter, B.K.: A & A **455**, 361 (2006)

Erdelyi, R., Fedun, V.: Sol. Phys. **238**, 41 (2006)

Erdelyi, R., Fedun, V.: Sol. Phys. **246**, 101 (2007)

Erdelyi, R.: In: Dwivedi, B.N., Narain U. (eds.) Physics of Sun and Its Atmosphere, p. 61. World Scientific, Singapore (2008)

Erdelyi, R., Fedun, V.: Sol. Phys. **263**, 63 (2010)

Evans, J.W., Michard, R.: ApJ **136**, 493 (1962)

Ferraro, V.C.A.: ApJ **119**, 407 (1954)

Finze, C.V.: The Sun—A User's Manual. Springer, Berlin (2008)

Flec, B., Deubner, F.L.: A & A **224**, 245 (1989)

Foukal, P.V.: Solar Astrophysics, 2nd edn. Wiley, New York (2004)

Foullon, C., Verwichte, E., Nakariakov, V.M., Fletcher, L.: A & A **440**, L 59 (2005)

Foullon, C., Verwichte, E., Nakariakov, V.M., Nykyri, K., Farrugia, C.J.: ApJ **729**, L8 (2011)

Frazier, E.N.: ApJ **152**, 557 (1968)

Gilbert, H.R., Thompson, B.J., Holzer, T.E., Burkepile, J.T.: AGU meet, SH12B - 0746 (2001)

Gilbert, H.R., Holzer, T.E.: ApJ **610**, 572 (2004)

Giovanelli, R.G.: Sol. Phys. **27**, 71 (1972)

Giovanelli, R.G.: Secrets of the Sun. Cambridge University Press, Cambridge (1984)

Goedbloed, J.P., Poedts, S.: Principles of MHD. Cambridge University Press, Cambridge (2004)

Goedbloed, J.P., Keppens, R., Poedts, S.: Advanced MHD. Cambridge University Press, Cambridge (2010)

Golub, L., Pasachoff, J.M.: The Solar Corona, 2nd edn. Cambridge University Press, Cambridge (2010)

Gomez, D.O., Dmitruk, P.A., Milano, J.L.: Sol. Phys. **195**, 299 (2000)

Goossens, M., Andries, J., Aschwanden, M.J.: A & A **394**, L 39 (2002)

Goossens, M., Andrias, J., Arregui, I., Doorselaere, T.V., Poedts, S.: AIP Conf. Proc. **784**, 114 (2005)

Goossens, M., Andries, J., Arregui, I.: Roy. Soc. Lond. Trans. **364**, 433 (2006)

Gough, D.G.: In: Zahn, J.P., Zinn-Justin, J. (eds.) Astrophysical Fluid Dynamics, vol. 399. Elsevier, Amsterdam (1993)

Gough, D.G.: Astrophys. Dynms., IAU Symp. **271**, 3 (2010)

Goutterbroze, P., Vial, J.C., Bocchialiani, J.W., Lemaire, P., Leibacher, J.W.: Sol. Phys. **184**, 253 (1999)

Gramer, L.: Geophys. Fluid Dyn. Lect. II **1** (2007)

Greiner, W.: Classical Electrodynamics. Springer, Berlin (1998)

Griffiths, D.J.: Introduction to Electrodynamics, 2nd edn. Prentice Hall of India, New Delhi (1994)

Gupta, G.R., Banerjee, D., Teriaca, L., Imada, S., Solanki, S.K.: 38th COSPAR Meeting, vol. 6, Bremen, Germany (2010)

Gurnett, D.A., Maggs, J.E., Gallagher, D.L., Kurth, W.S., Scarf, F.L.: J. Geophys. Res. **86**, 8833 (1981)

Handy, B.N., Deluca, E.E., McMullen, R.A., Schrijver, C.J., Tarbell, T.D., Title, A.M., Wolfram, C.J.: A & A Suppl. **193**, 1209 (1998)

Harvey, J., Hill, F., Kennedy, J., Leibacher, J.: In: Brown, T.M. (ed.) GONG 1992, vol. 397. Astrophysics Society of Pacific, San Francisco (1993)

Hasan, S.S., van Ballegooijen, A.A., Kalkofen, W., Steiner, O.: ApJ **631**, 1270 (2005)

Hasegawa, A.: Adv. Phys. **34**, 1 (1985)

Heyvaerts, J., Priest, E.R.: A & A **117**, 220 (1983)

Hillier, A., Berger, T., Isobe, H., Shibata, K.: ApJ **746**, 120 (2012)

Hollweg, J.V.: Adv. Space Res. **30**, 469 (2002)

Hudson, H., Davila, J.M., Dennis, B., Emslie, G., Lin, R., Ryan, J., Shore, G.: EAE **7939** (2003)

Hufbauer, K.: Exploring the Sun. Johns Hopkins University Press, Baltimore (1993)

Ichimoto, K., Suematsu, Y., Tsuneta, S., Katsukawa, Y., Shimizu, T., Shine, R.A., Tarbell, T.D., Title, A.M., Lites, B.W., Kubo, M., Nagata, S.: Science **318**, 1597 (2007)

Ireland, J., Priest, E.R.: Sol. Phys. **173**, 31 (1997)

Jackson, J.D.: Classical Electrodynamics, 2nd edn. Wiley, New York (1975)

Jain, R., Roberts, B.: Sol. Phys. **133**, 26 (1991)

Jensen, E., Orrall, F.Q.: ApJ **138**, 252 (1963)

Joarder, P.S., Nakariakov, V.M., Roberts, B.: Sol. Phys. **176**, 285 (1997)

Joarder, P.S., Narayanan, A.S.: A & A **359**, 1211 (2000)

Joarder, P.S., Nakariakov, V.M.: Geophys. Astrophys. Fluid Dynm. **100**, 59 (2006)

Joarder, P.S., Ghosh, S.K., Poria, S.: Geophys. Astrophys. Fluid Dynm. **103**, 89 (2009)

Jouve, L., Proctor, M.R.E., Lesur, G.: A & A **519**, A68 (2010)

Judge, P.G., Tarbell, T.D., Wilhelm, K.: ApJ **554**, 424 (2001)

Judge, P.: ASP Conf. Series. In: Leibacher, J., Stein, R.F., Uitenbrock, H. (eds): **354**, 259 (2006)

Kariyappa, R., Dame, L.: astroph doc arxiv 0804.3502K (2008)

Katsiyannis, A.C., Williams, D.R., McAteer, R.T.J., Gallager, P.T., Keenan, F.P., Murtagh, F.: A & A **406**, 709 (2003)

Khan, J.I., Aurass, H.: A & A **383**, 1018 (2002)

Khomenko, E., Centeno, R., Collados, M., Bueno, T.: ApJ **676**, L85 (2008)

Kippenhahn, R.: Discovering the Secrets of the Sun. Wiley, New York (1994)

Kliem, B., Dammasch, I.E., Kurdt, W., Wilhelm, K.: ApJ **568**, L 61 (2002)

Kosovichev, A.G.: arxiv:11o3.170v1 [astro-ph.SR] (2011)

Koutchmy, S., Zhughzhda, I.D., Locans, V.: A & A **120**, 185 (1983)

Kukhianidze, V., Zaqarashvili, T.V., Khutsishvili, E.: A & A **449**, L 35 (2006)

Ledoux, P.: Bull. Acad. Roy. Belg. College of Science, **48**, 240 (1962)

Lamb, H.: Proc. Roy. Soc. Lond. **7**, 122 (1909)

Lang, K.R.: Sun, Earth and Sky, 2nd edn. Springer, Berlin (2006)

Libbrecht, K.G., Woodard, M.F., Kaufman, J.M.: Ap. J. Suppl. Series, **74**, 1129 (1990)

Leibacher, J., Stein, R.F.: ApJ **7**, L191 (1971)

Leighton, R.B., Noyes, R.W., Simon, G.W.: ApJ **135**, 474 (1962)

Levich, E., Tzevetkov, E.: Phys. Rep. **128**, 1 (1985)

Liu, S., Zhang, H.Q., Su, J.T.: Sol. Phys. **270**, 89 (2011)

Malara, F., Primavera, L., Veltri, P.: Nonlin. Process. Geophys. **8**, 159 (2001)

Manners, J.: Static Fields and Potentials. Institute of Physics, Bristol (2000)

Mariska, T.J.: ApJ **620**, L 67 (2005)

Mariska, T.J.: ApJ **639**, 484 (2006)

Mathiridakis, M., Blookfield, D.S., Jess, D.B., Dhillon, V.S., Marsh, T.R.: A & A **456**, 323 (2006)

McAteer, R.T.J., Gallagher, P.T., Williams, D.R., Mathioudikas, M., Bloomfield, S.D., Philips, K.J.H., Keenan, F.P.: ApJ **587**, 806 (2003)

McIntosh, S.W., Bogdan, T.J., Cally, P.S., Carlsson, M., Hansteen, V.H., Judge, P.G., Lites, B.W., Peter, H., Rosenthal, C.S., Tarbell, T.D.: ApJ **543**, L 237 (2001)

McKenzie, J.F., Bond, R.A.B., Dougherty, M.K.: J. Geophy. Res. **92**, 1 (1987)

Mendoza-Briceno, C.A., Erdelyi, R., Leonardo Di G Sigalotti: ApJ **579**, L49 (2002)

Mendoza-Briceno, C.A., Leonardo Di G Sigalotti, Erdelyi, R.: Adv. Space Res **32**, 995 (2003)

Mendoza-Briceno, C.A., Erdelyi, R., Leonardo Di G Sigalotti: ApJ **624**, 1080 (2005)

Mendoza-Briceno, C.A., Erdelyi, R.: ApJ **648**, 722 (2006)

Meyer, F., Schmidt, H.: Z. Astrophys. **65**, 274 (1967)

Miles, A.J., Roberts, B.: Sol. Phys. **119**, 257 (1989)

Mitra-Kraev, U., Harra, L.K., Williams, D.R., Kraev, E.: A & A **436**, 1041 (2005)

Montesinos, B., Thomas, J.H.: Nature **390**, 485 (1997)

Moreton, G.E.: Astron. J. **65**, 494 (1960)

Moreton, G.E., Ramsey, H.E.: PASP **72**, 35 (1960)

Moses, D. et al.: Sol. Phys. **175**, 571 (1997)

Nakariakov, V.M., Roberts, B.: Sol. Phys. **159**, 213 (1995)

Nakariakov, V.M., Roberts, B., Mann, G.: A & A **311**, 311 (1996)

Nakariakov, V.M., Ofman, L., Deluca, E.E., Roberts, B.: Science **285**, 862 (1999)

Nakariakov, V.M., Ofman, L., Arber, T.: A & A **353**, 741 (2000)

Nakariakov, V.M., Verwichte, E., Berghmans, D., Robbrecht, E.: A & A **362**, 1151 (2000a)

Nakariakov, V.M., Ofman, L.: A & A **372**, L53 (2001)

Nakariakov, V.M.: IAU **203**, 353 (2001)

Nakariakov, V.M.: Solar MHD Notes. University of Warwick (2002)

Nakariakov, V.M.: In: Dwivedi, B.N. (ed.) Dynamic Sun. Cambridge University Press, Cambridge (2003)

Nakariakov, V.M., Melnikov, V.F., Reznikova, V.E.: A & A **412**, L 7 (2003)

Nakariakov, V.M., Verwichte, E.: Living Rev. Sol. Phys. **2**, 3 (2005)

Nakariakov, V.M.: ASP Conf. Ser. **369**, 221 (2007)

Narain, U., Sharma, R.K.: Sol. Phys. **181**, 287 (1998)

Narukage, N. et al.: A & A Suppl. **210**, 6304 (2007)

Nelson, J.G., Melrose, D.B.: Solar Radiophysics, p. 333. Cambridge University Press, Cambridge (1985)

Neupert, W.M.: Astron. J. **344**, 504 (1989)

Nitta, N.V., DeRosa, M.L.: ApJ **207**, L210 (2008)

Nocera, L., Ruderman, M.S.: A & A **340**, 287 (1998)

Ofman, L., Romoli, M., Noci, G., Kohl, J.L.: ApJ **491**, L 111 (1997)

Ofman, L., Klimchuck, J.A., Davila, J.M.: ApJ **493**, 474 (1998)

Ofman, L., Nakariakov, V.M., DeForest, C.E.: ApJ **514**, 441 (1999)

Ofman, L., Nakariakov, V.M., Sehgal, N.: ApJ **533**, 1071 (2000)

Ofman, L., Aschwanden, M.J.: ApJ **576**, L 153 (2002)

Ofman, L., Thompson, B.J.: arXiv: 1101 4249v2 (2011)

Oliver, R., Ballester, J.L.: Sol. Phys. **206**, 450 (2002)

O'Shea, E., Banerjee, D., Doyle, J.G.: A & A **463**, 7130 (2007)

Pandey, V.S., Dwivedi, B.N.: New Astron. **12**, 182 (2006)

Parker, E.N.: ApJ **128**, 664 (1958)

Parker, E.N.: Cosmical Magnetic Fields. Oxford University Press, Oxford (1979)

Payne-Scott, R., Yabsley, D.E., Bolton, J.G.: Nature **160**, 256 (1947)

Penderghast, K.H.: ApJ **123**, 498 (1956)

Priest, E.R.: Solar System Magnetic Fields. D. Reidel Publishers, Holland (1982)

Priest, E.R.: Solar MHD. D. Reidel Publishers, Holland (1982a)

Priest, E.R., Forbes, T.: Magnetohydrodynamics. IOP, Bristol (2000)

Pringle, J., King, A.: Astrophysical Flows. Cambridge University Press, Cambridge (2007)

Rae, I.C., Roberts, B.: A & A **119**, 28 (1983)

Ramesh, R., Subramanian, K.R., Sundararajan, M.S., Sastry, Ch.V.: Sol. Phys. **181**, 439 (1998)

Ramesh, R., Kathiravan, C., Satya Narayanan, A., Ebenezer, E.: A & A **400**, 753 (2003)

Ramesh, R., Satya Narayanan, A., Kathiravan, C., Sastry, Ch. V., Udaya Shankar, N.: A & A **431**, 353 (2005)

Ramesh, R., Kathiravan, C., Satya Narayanan, A.: Asian J. Phys. **13**, 277 (2004)

Ramesh, R., Kathiravan, C., Satya Narayanan, A.: ApJ **734**, 39 (2011)

Ramsey, H.E., Smith, F.S.: Astonomical J. **71**, 197 (1966)

Rhodes, E.J., Ulrich, R.K., Simon, G.W.: ApJ **218**, 901 (1977)

Roberts, B.: Sol. Phys. **69**, 27 (1981)

Roberts, B.: Sol. Phys. **69**, 39 (1981a)

Roberts, B., Edwin, P.M., Benz, A.O.: Nature, **385**, 688 (1983)

Roberts, B.: The Hydromagnetics of the Sun, p. 137. ESA, SP-220, Noordwijk, NL (1984)

Roberts, B., Edwin, P.M., Benz, A.O.: ApJ **279**, 857 (1984)

Roberts, B.: Plasma Physics College, ICTP, Italy, 1 (1990)

Roberts, B.: Geophys. Astrophys. Fluid Dynm. **62**, 83 (1991)

Roberts, B.: Sol. Phys. **193**, 139 (2000)

Roberts, B.: EASP **506**, 481 (2002)

Roberts, B.: EASP **547**, 1 (2004)

Roberts, B., Nakariakov. V.M.: Turbulence, Waves and Instabilities, NATO Adv. Research Workshop. In: Erdelyi, E., Petrovay, K., Roberts, B. (eds) (2003)

Roberts, P.H.: ApJ **122**, 508 (1955)

Rosenberg, R.L.: Sol. Phys. **15**, 72 (1970)

Rosenthal, C.S., Bogdan, T.J., Carlsoon, M., Dorch, S.B.F., Hanseen, V., McIntosh, S.W., McMurry, A., Nordlund, A., Stein, R.F.: ApJ **564**, 508 (2002)

Ruderman, M.S., Hollweg, J.V., Goossens, M.: Phys. Plasmas **4**, 75 (1997)

Rudraiah, N., Venkatachalappa, M.: J. Fluid Mech. **54**, 209 (1972)

Ruyotova, M.P.: Soviet Phys. JETP **94**, 138 (1988)

Satya Narayanan, A.: Ph.D. thesis, Indian Institute of Science, Bangalore (1982)

Satya Narayanan, A., Somasundaram, K.: Astrophys. Space. Sci. **109**, 357 (1985)

Satya Narayanan, A.: Il Nuovo Cimento **12**, 1121 (1990)

Satya Narayanan, A.: Plasma Phys. Contl. Fusion **33**, 333 (1991)

Satya Narayanan, A.: Phys. Scripta **47**, 800 (1993)

Satya Narayanan, A.: Phys. Scripta **53**, 638 (1996)

Satya Narayanan, A.: Bull. Astron. Soc. India **24**, 371 (1996a)

Satya Narayanan, A.: A & A **324**, 977 (1996)

Satya Narayanan, A., Ramesh, R.: Bull. Astron. Soc. India **30**, 573 (2002)

Satya Narayanan, A., Ramesh, R., Kathiravan, C., Ebenezer, E.: ASP Conf. Ser. **325**, 295 (2004)

Satya Narayanan, A., Ramesh, R., Kathiravan, C.: In: Singh, A.K. (ed.) Plasma 2004, p. 93. Allied Publishers, New Delhi (2004a)

Satya Narayanan, A., Ramesh, R.: In: Saha, S.K., Rastogi, V.K. (eds.) Magnetohydrodynamic Waves in the Solar Corona—A Review in the Book: Astrophysics of the 21st Century, p. 213. Anita Publications, New Delhi (2005)

Satya Narayanan, A., Pandey, V.S., Venkatakrishnan, P.: ESPM - 12. Freiburg, Germany (2008)

Satya Narayanan, A., Kathiravan, C., Ramesh, R.: IAU Symposium, **257**, 563 (2009)

Schmidt, H., Zirker, J.: ApJ **138**, 1310 (1963)

Schmidt, H., Stix, M.: Mitt. Astron. Gell. **32**, 182 (1973)

Schnack, D.D.: Lectures in Magnetohydrodynamics, Springer, Berlin (2009)

Schrijver, C.J., Hulbert, N.E., Engvold, O., Harvey, J.W.: Sol. Phys. **190**, 1 (1999)

Schrijver, C.J., Aschwanden, M.J., Title, A.M.: Sol. Phys. **206**, 69 (2002)

Schwartz, M.: Principles of Electrodynamics, McGraw Hill, New York (1972)

Singh, K.A.P., Dwivedi, B.N.: New Astron. **12**, 479 (2007)

Singh, A.P., Talwar, S.P.: Sol. Phys. **148**, 27 (1993)

Smerd, S.F.: NASA Symp. Goddard Space Flight Center 343 (1964)

Smeyers, P.: Annales de Astrophysique, **31**, 159 (1968)

Somasundaram, K., Uberoi, C.: Sol. Phys. **81**, 19 (1982)

Somasundaram, K., Satya Narayanan, A.: Plasma Phys. Contl. Fusion **29**, 497 (1987)

Sonnet, C.P., Giampapa, M.S., Mathews, M.S.: The Sun in Time. University of Arizona Press, Tucson (1991)

Spruit, H.C.: A & A **98**, 155 (1981)

Spruit, H.C., Roberts, B.: Nature **304**, 401 (1983)

Stein, R.F., Schwarz, R.: ApJ **131**, 442 (1972)

Stix, M.: A & A **4**, 189 (1970)

Strong, K.T., Saba, J.L.R., Haisch, B.M., Schmelz, J.T.: The Many Faces of the Sun. Springer-Verlag, Berlin (1998)

Suzuki, S., Dulk, G.A.: Solar Radiophysics. Cambridge University Press, Cambridge (1985)

Svestka, Z., Dennis, B.R., Woodgate, B.E., Pick, M., Raoult, A., Rapley, C.G., Stewart, R.T.: Sol. Phys. **80**, 143 (1982)

Tandberg-Hansen, E., Emslie, A.G.: The Physics of Solar Flares. Cambridge University Press, Cambridge (1988)

Taroyan, Y., Erdelyi, R.: Sol. Phys. **251**, 523 (2008)

Thomas, J.H.: In: Cram, L.E., Thomas, J.H. (eds.) The Physics of Sunspots. Sac Peak Observatory, Sunspot, New Mexico (1981)

Thomas, J.H.: Ann. Rev. Fluid Mech. **15**, 321 (1983)

Thomas, J.H.: Proc. MPA/LPARL Workshop, Ed. Schmidt, H.U., Sept 16–18, 126 (1985)

Thomas, J.H., Weiss, N.O.: Sunspots and Starspots. Cambridge University Press, Cambridge (2008)

Thomas, J.H.: In: Hasan, S.S., Rutten, R.J. (eds.) Astrophysics Space Science Proceeding, p. 229. Springer, Berlin (2010)

Thompson, B.J., Plunkett, S.P., Gurman, J.B., Newmark, J.S., St. Cyr, O.C., Michels, D.J.: Geophys. Res. Lett., **25**, 246 (1998)

Tirry, W.J., Cadez, V.M., Erdelyi, R., Goossens, M.: A & A **332**, 786 (1998)

Tskilauri, D., Nakariakov, V.M.: A & A **379**, 1106 (2001)

Uberoi, C., Somasundaram, K.: Plasma Phys. **22**, 747 (1982)

Uberoi, C., Somasundaram, K.: Phys. Rev. Lett. **49**, 39 (1982a)

Uberoi, C., Satya Narayanan, A.: Plasma Phys. Contl. Fusion **28**, 1635 (1986)

Uchida, Y.: Sol. Phys. **4**, 30 (1968)

Uchida, Y.: PASJ, **22**, 341 (1970)

Uchida, Y.: In: Proc. Symp. GSFC, Greenbelt, M.D. Ramaty, R., Stone, R.G. (ed.) **342**, 577 (1973)

Uchida, Y.: Sol. Phys. **39**, 431 (1974)

Uchida, Y., Sakurai, T.: PASJ **27**, 259 (1975)

Ulrich, R.: ApJ **162**, 993 (1970)

Ulaby, F.T.: Fundamentals of Applied Electromagnetism, Prentice Hall International Inc. London (1997)

Ulmschneider, P.: In: Antia, H.M., Bhatnagar, A., Ulmschneider, P. (eds.) Lectures on Solar Physics, vol. 619, p. 232. Springer, Berlin (2003)

van Ballegooijen, A.A.: ApJ **304**, 828 (1986)

Vandakurov, Ya. V.: ApJ, **149**, 435 (1967)

Varga, E., Erdelyi, R.: ESA SP **464**, 255 (2001)

Varga, E., Erdelyi, R.: PADEU **11**, 83 (2001a)

Vigeesh, G., Hasan, S.S., Steiner, O.: A & A **508**, 951 (2009)

Vrasnak, B., Magdalemic, J., Aurass, H., Mann, G.: EASP **506**, 409 (2002)

Walsh, R.: In: Hanslmeier, A., Messerotti, M., Veronis, A. (eds.) The Dynamic Sun: Proceedings of the Summer School and Workshop, Austria, p. 187. Kluwer, Dordecht (1999)

Wang, T.J., Solanki, S.K., Curdt, W., Innes, D.E., Dammasch, I.E.: ApJ **574**, L 101 (2002)

Wang, T.J., Solanki, S.K., Curdt, K., Innes, D.E., Dammasch, I.E., Kleim, B.: A & A, **406**, 105 (2003a)

Wang, T.J., Solanki, S.K., Innes, D.E., Curdt, K., Marsch, E.: A & A, **402**, L17 (2003b)

Wang, T.J.: ESA SP **547**, 417 (2004)

Wang, T.J.: Astroph – 12605 (2006)

Warmuth, A., Vrasnak, B., Magdalemic, J., Hanslmeir, A., Otruba, W.: A & A **410**, 1101 (2004)

Warmuth, A., Mann, G., Aurass, H.: ApJ **626**, L 121 (2005)

Weatherall, J.C., Goldman, M.V., Nicholson, D.R.: ApJ **246**, 306 (1981)

Wentzel, D.G.: ApJ **227**, 319 (1979)

White, S.M., Thompson, B.J.: ApJ **620**, L 63 (2005)

Wikstol, O., Hansteen, V.H., Carlsson, M., Judge, P.G.: ApJ **531**, 1150 (2000)

Wild, J.P., McCreaky, L.L.: Australian J. Sci. Res. **3**, 387 (1950)

Wild, J.P., Smerd, S.F., Weiss, A.A.: Ann. Rev. Astro. Astrophys. **1**, 291 (1963)

Williams, D.R., Phillips, K.J.H., Rudway, P., Mathioudakis, M., Gallager, P.T., O'Shea, E., Keenan, F.P., Read, P., Rompolt, B.: MNRAS **326**, 428 (2001)

Williams, D.R., Phillips, K.J.H., Rudway, P., Mathioudakis, M., Gallager, P.T., O'Shea, E., Keenan, F.P., Read, P., Rompolt, B.: MNRAS **336**, 747 (2002)

Wolf, C.: ApJ **177**, L87 (1972)

Xiong, D.R., Deng, L.C.: Acta Astron Sinica **50**, 349 (2009)

Yan, Y.: Coronal and stellar mass ejections. Proceedings IAU Symposium No. 226, p. 277 (2005)

Zaqarishvili, T.V., Khutsishvili, E., Kukhianidze, V., Ramishvili, G.: A & A **474**, 6 (2007)
Zhukov, V.I.: Sol. Phys. **139**, 201 (1992)
Zlotnik, Ya, E., Zaitzev, V.V., Aurass, H., Mann, G., Hoffman, A.: A & A **410**, 1011 (2003a)
Zlotnik, Ya, E., Zaitsev, V., Aurass, H., Mann, G. : Adv. Space Res. **35**, 1774 (2003b)
Zukhov, V.I.: Sol. Phys. **173**, 15 (1997)

Index

A. Satya Narayanan, *An Introduction to Waves and Oscillations in the Sun*, Astronomy
and Astrophysics Library, DOI 10.1007/978-1-4614-4400-8,
© Springer Science+Business Media New York 2013

Printed in the United States
By Bookmasters